Pharmaceutical Calibration, Validation and Qualification: A Comprehensive Approach

Shiv Shankar Shukla • Ravindra Kumar Pandey •
Beena Gidwani • Gunjan Kalyani

Pharmaceutical Calibration, Validation and Qualification: A Comprehensive Approach

Springer

Shiv Shankar Shukla
Columbia Institute of Pharmacy
Raipur, Chhattisgarh, India

Ravindra Kumar Pandey
Columbia Institute of Pharmacy
Raipur, Chhattisgarh, India

Beena Gidwani
Columbia Institute of Pharmacy
Raipur, Chhattisgarh, India

Gunjan Kalyani
Columbia Institute of Pharmacy
Raipur, Chhattisgarh, India

ISBN 978-981-19-9001-4 ISBN 978-981-19-9002-1 (eBook)
https://doi.org/10.1007/978-981-19-9002-1

This Springer imprint is published by the registered company Springer Nature Singapore Pte Ltd.
The registered company address is: 152 Beach Road, #21-01/04 Gateway East, Singapore 189721, Singapore

"The best parts of your father are the best parts of you. Never forget where you came from".

Late Shri R. S. Shukla

If I could write a story, it would be the greatest ever told.

I'd write about my daddy, for he had a heart of gold.

My dad, he was no hero, known around this world.

He was everything to me.

Prof. Shiv Shankar Shukla

Preface

I am delighted to write the preface for the first edition of *Pharmaceutical Calibration, Validation and Qualification: A Comprehensive Approach*.

The uniqueness of the book can be accounted with the multitude of its content which will encompass diverse topics under one cover which no other title has done so far. There is complete and exhaustive coverage of the topics pertaining to validation which makes the book indispensable for readers.

Various chapters mentioned in the content will cover detailed aspect of the respective topic with respect to the advancements in the field; this will arouse the interest of the reader which will facilitate continued reading for students, academicians, researchers, industry persons as well.

This book includes the detailed description of the topic "Validation of herbals" which is an essential part with respect to the current scenario as majority of world's population relies on herbal medicines.

In a nutshell, the book will serve as a complete reference material for the topic under one cover which will make the understanding easy and interactive.

This book will be a backbone understanding for the validation concept.

Sincere thanks to our publishers for their kind gesture and cooperation.

Highly indebted to the co-authors for their generous help, cooperation, and criticism.

Note of thanks to the Columbia Institute of Pharmacy for providing the platform for stand up.

Earnest thanks to the Management, Shri Kishore Jadwani, Chairman, Jan Pragati Education Society (JPES) Raipur, and Shri Harjeet Singh Hura, Secretary, Jan Pragati Education Society (JPES) Raipur, for their active help and cooperation.

Raipur, Chhattisgarh, India — Shiv Shankar Shukla
October 2022

Contents

Part II Qualification

About the Editors

Shiv Shankar Shukla gained Ph.D. from the University Institute of Pharmacy, Pt. Ravi Shankar Shukla University. Presently, he is Professor at Columbia Institute of Pharmacy. He has more than 93 publications to his credit in reputed journals. His thrust area is method development, chemical fingerprinting, and pharmaceutical analysis. He has authored three books, *Inflammation: Natural Resources and Its Applications* in Series: Springer Briefs in Immunology, *Finger Printing Analysis & Quality Control Methods of Herbal Medicines* in CRC Press Taylor & Francis Group, United Kingdom, and *Insight on Genotoxicity* in CRC Press Taylor & Francis Group, United Kingdom. He has written various book chapters published in books from reputed publishing houses. Serving as Co-Coordinator of DST-FIST Grant (Fund for Improvement in Science and Technology Infrastructure) by the Department of Science and Technology, Government of India.

Ravindra Kumar Pandey gained Ph.D. from the University Institute of Pharmacy, Pt. Ravi Shankar Shukla University. Presently, he is Professor at Columbia Institute of Pharmacy. He has more than 87 publications to his credit in reputed journals. He has authored three books, *Inflammation: Natural Resources and Its Applications* in Series: Springer Briefs in Immunology, *Finger Printing Analysis & Quality Control Methods of Herbal Medicines* in CRC Press Taylor & Francis Group, United Kingdom, and *Insight on Genotoxicity* in CRC Press Taylor & Francis Group, United Kingdom. He has written various book chapters in reputed

Publishing House. Serving as Co-Coordinator of DST-FIST Grant (Fund for Improvement in Science and Technology Infrastructure) by the Department of Science and Technology, Government of India.

Beena Gidwani did her Post Doctoral under the Scheme of DST-SERB-NPDF and Ph.D. from the University Institute of Pharmacy, Pt. Ravishankar Shukla University, Raipur (C.G.). Presently, she is serving as an Associate Professor at the Columbia Institute of Pharmacy. She has more than 60 research and review publications to her credit. She has authored two books, *Techniques for preparation of host-guest inclusion complexes: Techniques of Complexation* in LAP LAMBERT Academic Publishing and *Insight on Genotoxicity* in CRC Press Taylor & Francis Group, United Kingdom, and two book chapters in reputed publishing houses.

Gunjan Kalyani earned his post graduate studies from the Chhattisgarh Swami Vivekanand Technical University (CSVTU), Bhilai, Chhattisgarh. Presently, he is an Assistant Professor at Columbia Institute of Pharmacy. He has 10 years of research and academic experience. His research area of interest includes pharmaceutical analysis, ethnomedicine, and medical anthropology. He has more than 30 publications to his credit in various journals of international and national repute. He has authored one book with the title *Insight on Genotoxicity* in CRC Press Taylor & Francis Group, United Kingdom, and two book chapters in reputed publishing houses.

Part I
Calibration

Chapter 1
Introduction to Calibration

Abstract In all areas of science, business, and industry, as well as in daily life, measurement is essential. Measurement calls for the employment of a measuring device. In order to establish the link between the values of quantities indicated by measuring equipment and systems under predetermined conditions, calibration is a collection of techniques. The calibration process includes comparing the device to main or secondary standards. In essence, calibration is the comparison of a higher standard, traceable to an international or national norm, or a recognized alternative, under predetermined circumstances. This chapter's objective is to examine the notion of calibrating, with an emphasis on the prerequisites, advantages, and method. By comparing a measuring standard or instrument to higher standards one at a time until it achieves an accepted norm on a national or international scale, traceability is an essential component in determining the accurate calibration of the standard or instrument. Thanks to technological advancements, there are presently five different types of traditional calibration procedures: electro-technical calibration, non-electrical calibration, on-site calibration, e-calibration, and intelligent/self-calibrating devices and sensors. For all instruments and equipment, the four crucial parameters—flow, sensor, images, and food items—must be calibrated in accordance with these categories. Therefore, calibration is essential whenever measurements are important because it enables individuals and organizations to have confidence in the results they monitor, record, and subsequently manage.

Keywords Calibration · Measurement · Accuracy · Traceability · Equipment

S. S. Shukla et al., *Pharmaceutical Calibration, Validation and Qualification: A Comprehensive Approach*, https://doi.org/10.1007/978-981-19-9002-1_1

Introduction

The process of giving physical objects numerical values or quantification is referred to as measurement. This makes it easier to establish connections between them in terms of a particular physical attribute. In all areas of science, business, and industry, as well as in daily life, measurement is essential. Measurement calls for the employment of a measuring device. All measuring equipment must be calibrated in order to guarantee their dependability and accuracy. Instrument calibration aims to remove or minimize bias from an instrument's readings across a broad range of continuous values. To compare results to accepted standards or references, instruments, and related procedures need to be calibrated [1].

Definition

The process of calibrating involves comparing a measuring instrument to a reference standard in order to ascertain its precision. In order to bring the instrument into compliance with the standard, it could also need to be adjusted. Precision and accuracy, the two most important parameters of the measurement equipment, are analysed and modified during calibration [1].

A collection of procedures known as calibration is used to establish the relationship between the known values of a measurand (the parameter that is being measured), and the values given by a measuring device, system, or material measure under certain circumstances, according to the International Organization for Standardization Publication International Vocabulary of Basic and General Terms in Metrology [1].

The measurand is the term used to describe the quantity being measured. It is described as a set of rules or requirements (not a numerical value). The relevant influence values are specified by the measurand, and they must be detailed in such detail that any uncertainty has no effect on the calibration's required precision [1].

Calibration mainly is done to realize the following two purposes:

- determination of the accuracy of the measured data
- providing traceability to the measurement

Purpose of Calibration [2] (Fig. 1.1)

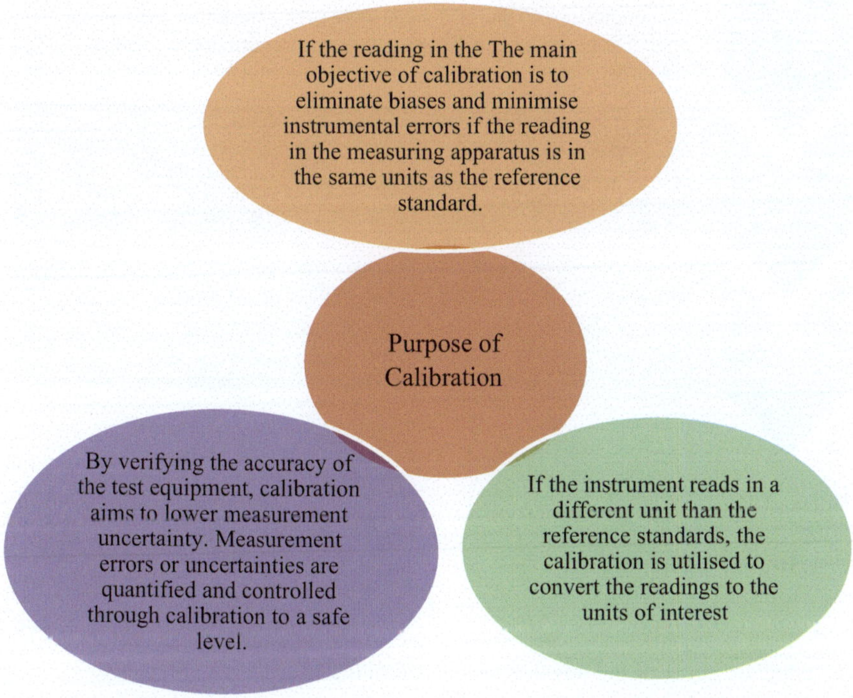

Fig. 1.1 Purpose of calibration

Need of Calibration (Fig. 1.2)

Thus, calibration ensures that a measuring device's value is always accurate and trustworthy. Therefore, calibration is an essential step in every measurement process [2].

Need of Calibration

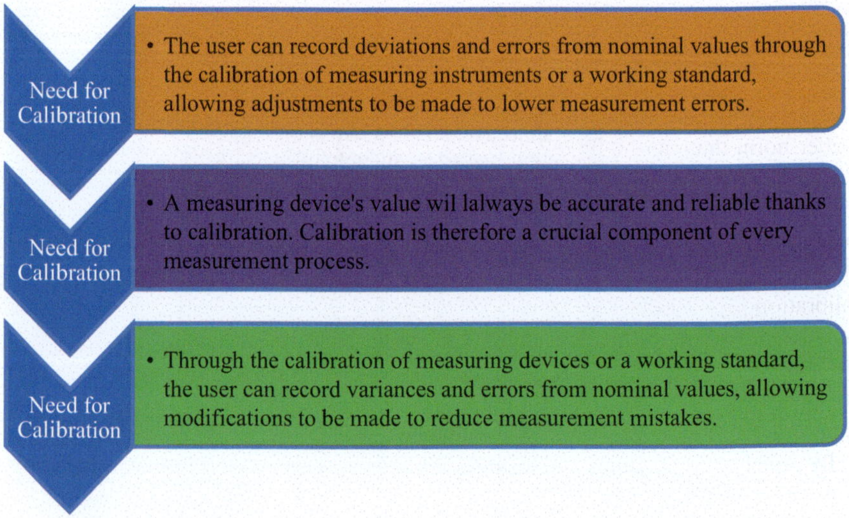

Fig. 1.2 Need of calibration

Benefits of Calibration

Calibration is regarded as a necessary stage in the operation of instruments. The following are some of the advantages of calibrating [3].

It ensures uniformity and compatibility, results in assessments of the instruments' repeatability and reproducibility, boosts productivity by guaranteeing accurate measurements, and makes it easier to apply relevant requirements. It establishes whether measurements taken before the calibration were accurate, ensures consistency and compatibility, results in repeatability and reproducibility assessments of the instruments and processes, increases efficiency by ensuring accurate measurements, facilitates the implementation of applicable rules and legislation that regulate the use of equipment in a specific application, and produces documentation of instrument and process performance to satisfy regulatory requirements such as ISO 9000, ISO 1400, and QS-9000.

Without calibration, product quality would suffer, leading to low productivity, legal troubles, and high rates of product failure, all of which will raise maintenance and associated costs. Regular instrument calibrations can boost performance reliability by showing a graphical picture of the equipment's uncertainty over time. This will make it easier to assess how well the instrument or life product is working and provide more precise replacement and depreciation forecasts. Measurements made in conformity with global standards boost their acceptance globally and, as a result, their competitiveness. As technology advances and test and measuring instrument laws and legislation change, calibration helps ensure the compliance and validity of measurements and procedures under changing situations [3].

Traceability

Traceability is the idea that a measuring standard or instrument may be properly calibrated by comparing it to higher standards one at a time until it reaches an acceptable national or global norm. In essence, calibration is a comparison to a higher norm that can be linked to a national, international, or accepted alternative standard. To make sure the measurement is accurate and reproducible, two or three measurements of the same parameter are typically compared. A measurement needs to be traceable to a recognized standard in order to be compared. The physical unit of measurement should be able to be traced back to the ultimate basic unit through calibration [3].

Greater precision/uncertainty is indicated by a higher standard (10:1 ratio is preferable ratio, but ideally it should not be less than 3:1.) An increased resolution (ideally, the resolution should be increased by a factor of 10:1, 5:1, or 2:1). The ratio indicated above is the test accuracy ratio. The ratio should be as high as is economically practical because the risk of making a poor measurement judgement decreases as the ratio increases [3].

Test Accuracy Ratio (TAR)

Uncertainty of unit under calibration.

$$\text{TAR} = \text{Uncertainty of the standard}$$

TAR is currently usually referred to as the test uncertainty ratio because accuracy is considered a qualitative rather than a quantitative requirement. As a result, the term uncertainty is used to define this characteristic rather than specifying the precision of a measuring device [4].

Types of Calibration

The traditional calibration procedure is changing on at least three fronts as a result of the availability of advanced technology as follows:

- electro-technical/electronic calibration
- non-electrical calibration
- on-site calibration
- e-calibration via the Internet and other communication methods
- instruments and sensors that are intelligent and self-calibrating

Parameters to be Calibrated

The properties of the testing procedure or measuring apparatus establish the physical qualities that need to be calibrated. Numerous measurement tools and procedures are accurate, thanks to calibration [4] (Fig. 1.3).

- *Flow calibration:* There is a wide range of equipment and facilities available for measuring the flow of liquid, air, or solids. Following the specification of the measuring procedure by a suitable flow metre setup, static or dynamic calibrations can be carried out. A reservoir, a pumping system, a pipeline, a flow metre placed on the pipeline for testing, a collecting system, computers and interfaces, supporting software, and other components may be found in a calibration facility for static-gravimetric liquid flow. You can calibrate the flow of fluid via the metre by gathering the necessary mass of steadily flowing fluid over a time period that is measured.
- *Sensor calibration*: Sensors need to be calibrated once their data has been combined with a signal conditioning system. A known input signal needs to be introduced into the sensor in order to calibrate the processing component of the system. By looking at the output, a suitable output scale for that specific application can be configured. Dynamic calibration is needed when the sensor is used to measure time-varying inputs. Most of the time, the sensor's top step response's transient behaviour can be used to determine its dynamic response.
- *Calibration of food products*: Device calibration becomes more challenging if food contains more chemical components. The following parameters must be determined in the case of honey using calibrated tools and methods: glucose, maltose, turanose, acidity, moisture content, and other sugars.

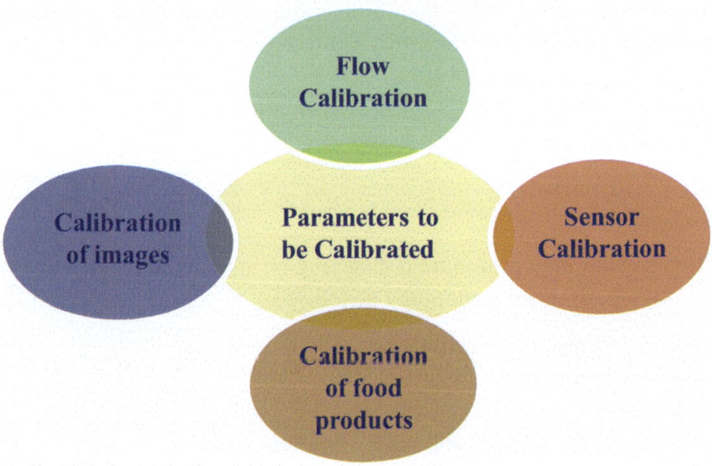

Fig. 1.3 Parameters to be calibrated

- *Calibration of images*: Calibration is one of the first steps in image processing. For example, astronomical images are calibrated to eliminate camera, light pollution, and distortion effects. There are several techniques used to ensure calibrated images, including reducing thermal, readout, and other influences. Black frames are utilized to offset the noise produced by the camera electronics, and the cameras are cooled below specific temperatures for thermal effects.

Calibration attempts to eliminate or minimize bias in the measuring system relative to the reference base in accordance with a predetermined algorithm. Regardless of the direction of the measurement, the bias may be created by the instrument itself. Additionally, linear drift across the measurement period has the potential to cause bias [4].

Calibration Techniques and Methods

In general, there are four methods for calibrating an instrument or piece of equipment (Fig. 1.4):

- spanning
- nulling
- zeroing
- linearization

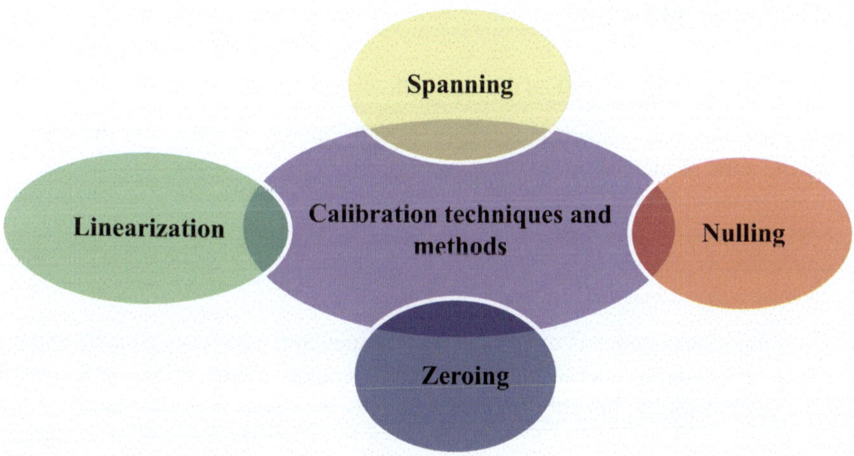

Fig. 1.4 Calibration techniques and methods

Nulling

- Using this technique, one can check whether the nominal value of UUC and the reference are identical. Instruments that operate on this tenet include comparators, null detectors, and galvanometers.
- The scale displays how sensitively these metres can be detected. Since the Null process is sensitive to even slight variations in known and unknown variables, nulling is used for more exact measurements.
- The detection sensitivity of null detectors is dependent on the scale of the indicator, and they do not require calibration for the observed values.

Zeroing
Device-to-device offset errors, variance (trim errors), mechanical stresses (mounting stresses), temperature changes, and ageing shifts are all taken into account.

Spanning
It is simply calibrating the instrument to read the known concentration of a gas, substance, or chemical; it is the parameter range that the apparatus can measure.

Linearization
To approximate a nonlinear system with a linear one is the aim of linearization. For instance, linearization is used to convert an NTC thermistor's output of resistance to voltage when it is connected to a computerized data acquisition system [5].

Methods of Calibration

Calibration can be accomplished through following methods:

- *Standard method*—Direct, indirect, ratio, substitution, transfer, and differential are the sorts of method.
- *Non-standard method.*
- *Laboratory developed method.*
- *Expansion of scope.*
- *Customer given method.*

The validation parameters to be used for non-standard, laboratory created, expansion of scope, and customer technique include selectivity, range, robustness, repeatability, reproducibility, linearity, LOD, LOQ, bias, precision, and uncertainty of the results [5].

Calibration in Analytical Measurements

In analytical chemistry, calibration has two components:

- calibration of measuring instruments
- calibration of the analytical method

Similar to how volumetric flasks and other measuring devices are calibrated and measurement traceability to SI units is created, so too are thermometers. The value provided by measuring equipment during an analytical process could be, for instance,

- the integral of a peak for a high-pressure liquid chromatograph in any units
- the spectrophotometer's optical density for atomic absorption
- the strength of the current that a flame photometer generates [5]

Process of Calibration

A recorded, confirmed, and validated method that describes a number of steps that adhere to a particular technique is called a calibration procedure. A hierarchical process is used to carry out the calibrating process [5].

- The primary reference standard is given a value based on a direct comparison to the SI reference base at the most fundamental level. With values that are independent of other standards of the same amount, fundamental standards are acknowledged as having the highest metrological quantities [5].
- By comparing the secondary reference standards to primary standards of the same quantity and making the necessary corrections using a high precision comparator, the secondary reference standards are calibrated at the second level [5]
- To calibrate instruments and procedures against secondary reference standards or their equivalents at the third level, working standards are utilized. The generation of genuine signals as well as a variety of other activities, such as finding mathematical relationships between impact values and sensor indicators, may be included in the calibration process. The most important calibration consideration is the relationship between a single measurement and the unit's reference base, which serves as the main authority [5].

A Successful Calibration Procedure Requires the Following Basic Steps

- Choosing a reference standard that fits the range of interest and has values that are known.
- Utilizing the device or the benchmark.
- Establishing the link between the measured and known values of the reference standard using calibration curves (i.e. least-squares fit).
- Measurement correction by the use of calibration curves.

- Putting together adequate records of the calibration process, findings, and analysis.
- Calibration curves are used to adjust measurements.

Both static and dynamic conditions can be used to calibrate instruments and procedures. If there are several inputs and outputs for the calibrated variable, the input is modified in steps over a predetermined range in both increasing and decreasing directions. The observable output is hence a function of that single input. A family of relationships between the inputs and outputs can be created by repeating this process with different inputs for improved outcomes [6].

In multivariable situations, the input/output relationship usually demonstrates statistical characteristics. These characteristics allow for the creation of calibration curves that are suitable for statistical analysis [6].

The readings from the test item are compared to the reference standards throughout the calibration process so that values can be assigned based on the known values of the reference standard. Reference standards can include resistors, length standards, voltage standards, and more. However, comprehensive calibration may not always be possible due to biases in the measurement and instrument, as well as unanticipated random mistakes [6].

Mathematically:

$$\text{Ideal value} = \text{Measured value} + \text{Bias} + \text{Error}$$

Similarly, a reference value may be subject to bias and error:

$$\text{Ideal reference value} = \text{Reference value} + \text{Bias} + \text{Error}$$

This leads to a deficiency in the calibration:

$$\text{Deficiency} = \text{Ideal measured value} - \text{Ideal reference value}$$

Five reference standards are needed for a linear calibration curve, while ten reference standards are sufficient for more complicated calibration models. The majority of calibrations today use computers to record and analyse data. Software tools may be used to aid in the analysis of the data once the results have been obtained. The least-squares method is used by the majority of packages to estimate the coefficients. Several of the software may perform a weighted fit, if the measurement errors are not constant across the calibration interval. Data such as calibration curve coefficients, standard deviations, residual standard deviation of the fit, and goodness of fit are displayed by the software tools [6].

Calibration Personnel

The calibration is completed with the aid of qualified personnel using the appropriate instruments to reflect the reference standards. National authorities and enterprises maintain adequate equipment as operating standards and secondary requirements. Standards, laws, and procedures that have been developed regulate the calibration process [7].

The calibrating people make the following assumptions during the calibration:

- The handling methods are standard.
- In the testing environment, the reference standard and the test item perform identically.
- The measurement's random mistakes are unrelated to one another.
- A distribution curve with the same standard deviation is produced by many measurements.
- Throughout the measurement, the reference standards and the test item remain constant.
- Because bias exists in both the test item and the reference standard, it can be eliminated by comparing the two readings' differences.
- The gap should be zero after the calibrations or understandable from relevant graphs and statistical connections.
- Calibration can be repeated and produces accurate results.

To ensure the accuracy of the calibration, a qualified calibration staff is required. For safety reasons and to avoid contamination, workers who perform calibrations may need to dress correctly (static-free clothing, gloves, face masks, etc.). Different operators may give measurements with a range of magnitudes and indications while having significant training and experience. Measurements taken by several operators could be plotted and contrasted to solve this issue. Another option is for the same operator to maintain multiple calibration curves. In automatic calibrations, this might not be an issue [7].

Calibration Laboratory

A controlled environment should be used for calibration in the first, second, or third party laboratory. To ensure precise traceability, it should be done at a calibration lab that has been accredited or in the lab of the manufacturer or supplier. Technical specifications and parameters for calibration laboratories should adhere to ISO/IEC 17025 [7].

The essential requirements for a calibration laboratory are depicted in the diagram below (Fig. 1.5).

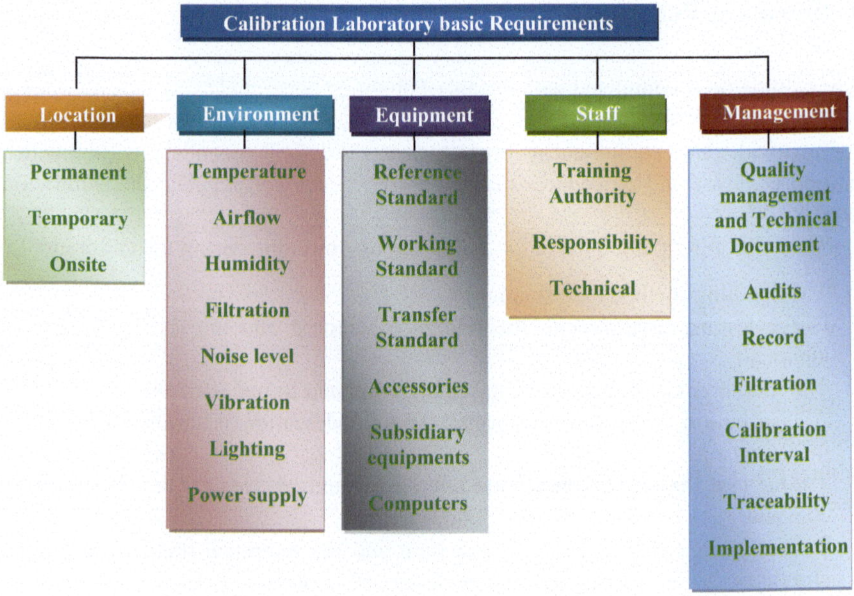

Fig. 1.5 Basic requirement of calibration laboratory

Calibration Certificate and Report

Review and evaluation of the calibration certificate should be done as described below before using the instrument for measurement. Such tests should be carried out when acquiring a certificate for an instrument that has been purchased or calibrated before. First, the certificate needs to work with the instrument. This is confirmed by contrasting the serial or identification number on the instrument with the number on the certificate. The calibration certificate should be read over and evaluated in the manner described below before using the equipment for measurement. Such tests should be carried out when acquiring a certificate for an instrument that has been purchased or calibrated before. First, the certificate needs to work with the instrument. This is confirmed by contrasting the serial or identification number on the instrument with the number on the certificate [8].

The criteria under which the calibration results are valid must be stated in the calibration documentation, or calibration report. These conditions, which we'll refer to as the calibration validity conditions, encompass the values (or range of values) of all important influence quantities for which the calibration results are valid [8].

Recalibration

When compared to a traceable reference, calibration ensures that the value provided by a measuring device is precise and repeatable. However, it cannot be assumed that the instrument will always deliver precise and repeatable readings after it has been calibrated. Therefore, instrument recalibration is necessary to determine how far off from a reference value or standard the instrument is, as well as to ensure that the variation is adequate for the measurement over time [8].

Recalibration is also necessary under the following two conditions:

- when the instrument undergoes routine maintenance
- when the instrument goes out of order and is repaired

While frequent calibration would accomplish the stated objective, it must be considered that calibration comes at a large cost. Therefore, selecting the appropriate calibration frequency, which strikes a balance between risk and cost, becomes an essential measuring task [8].

Factors Influencing Recalibration

The time between calibrations of measuring equipment is influenced by a number of factors. The important factors are:

- manufacturer's recommendations
- the volume and intensity of usage
- the measurement's accuracy and precision
- the possibility that a measuring device would exceed its tolerance while being used
- equipment kind, wear and drift propensity, and manufacturer's recommendations
- circumstances of the environment, such as temperature, vibration, radiation, etc.
- data trends derived from historical calibration records
- history of maintenance and services
- frequency of cross-checking against other measurement tools or reference standards
- handling, storage, and risk management practices
- training level of serving personnel

The purpose of a periodic calibration, therefore, is:

- to calculate the measurement uncertainty related to the measurement instrument's or reference standard's departure from a reference value
- to reassure that the reference standard or measuring equipment can still be used to achieve the measurement uncertainty

While frequent calibration would accomplish the stated objective, it is important to consider the substantial cost of calibration. Therefore, selecting the right calibration frequency—which strikes a balance between cost and risk—becomes an essential measuring task. Techniques for assessing the interval must be created after the first calibration has been established to ensure that neither the danger of the instrument being out of calibration nor the cost of doing so rises. Several of these processes have been documented in international standards. Two often utilized techniques are the control chart method and the staircase or calendar-time approach [8].

Conclusion

Calibration is a critical component of any organization's performance and growth in the pharmaceutical sector and laboratories. Instrument, equipment, process, and related services calibration should be pre-planned and scheduled according to a defined protocol. In the calibration process, accuracy and dependability are critical. For a calibration laboratory, there are a few basic requirements that must be met. Calibration is always done in a controlled environment in the I/II/III party laboratory. (Only to the extent necessary, the environment must be managed.) It should be carried out at an accredited calibration laboratory, as well as a manufacturer's or supplier's laboratory, to ensure accurate traceability. Inspection, measuring, and testing of equipment are all common measurement characteristics (IM&TE). To calibrate any equipment, a known amount of the variable to be measured must be generated and applied to the unit under test. The four essential factors that must be calibrated in any equipment are temperature, humidity, pressure, and mass. The frequency of calibration, which includes recalibration, is also a requirement. As a result, calibration is critical wherever quantitative measurements are required, as it ensures accuracy.

References

1. American National Standards Institute/National Conference of Standards Laboratories (1994) Calibration laboratories and measuring and test equipment: general requirements. National Conference of Standards Laboratories, Boulder, Colorado. (ANSI/NCSL Z540–1-1994)
2. Eren H (2005) Calibration process. Hand book of measuring system design. John Wiley & Sons, Ltd., Chichester, UK, pp 271–277. ISBN: 0-470-02143-8
3. Phillips SD, Estler WT, Doiron T, Eberhardt KR, Levenson MS (2001) A careful consideration of the calibration concept. J Res Natl Inst Stand Technol 106:371–379
4. ISPE (2002) GAMP good practice guide: calibration Management. International Society of Pharmaceutical Engineers, Tampa, FL
5. www.calibrationlaboratory.in. Accessed 26 Jan 2022
6. www.stqc.gov.in. Accessed 26 Jan 2022
7. www.rscalibration.com. Accessed 26 Jan 2022
8. Guide, I. S. O (1990) 25, General Requirements for the Competence of Calibration and Testing Laboratories. International Organization for Standardization, Geneva, Switzerland

Chapter 2
Errors and Uncertainties in Calibration

Abstract When measuring a physical quantity, it is required to provide a quantifiable indicator of the result's quality so that others who use it may assess its dependability. Measurement results cannot be compared without this information, either among themselves or with reference values provided in a specification or standard. The disparity between the true and observed values is defined as an error. Every measurement has an unknown and unknowable inaccuracy. "Measurement uncertainty" is the term for this unknown inaccuracy. Systematic errors are errors that are constant and repeatable. Random error is a term used to describe the scatter of repeated measurements around a mean value. The magnitude and sign of the inaccuracy varies at random and cannot be predicted precisely. They are usually easy to spot and are always present in a measurement. The smallest change or increment in the measured amount that the instrument can detect is formally defined as resolution. The number of digits displayed on the output of digital equipment is frequently connected with resolution. Calibration of instruments can be static or dynamic. The three principal sources of uncertainty are the reference utilized (measurement standard), the unit under calibration (UUC) (measurement result), and the calibration technique employed (under defined parameters). The uncertainty in a measurement's result is usually made up of multiple components that can be classified into two groups based on how their numerical value is calculated; these are of two types: Type A—Statistical procedures, such as average and standard deviation, are used to evaluate them. Type B includes those that are evaluated through other means and are based on scientific judgement using all available information, such as previous measurement data, manufacturer's specifications, data provided in calibration certificates, and so on. As a result, uncertainty can be evaluated and reported based on the data, type, and method that can be investigated quantitatively.

Keywords Errors · Uncertainty · Calibration · Measurand · Specification · Reference

S. S. Shukla et al., *Pharmaceutical Calibration, Validation and Qualification: A Comprehensive Approach*, https://doi.org/10.1007/978-981-19-9002-1_2

Introduction

Quantity measurement is based on a set of international fundamental standards. Others are derived from these fundamental norms, which are fully accurate. Any departure from the actual value is defined as an error in the measurement of a physical quantity. When reporting the result of a physical quantity measurement, it is required to provide a quantitative indicator of the result's quality so that others who use it can judge its dependability. Measurement results cannot be compared without this information, either among themselves or with reference values provided in a specification or standard. As a result, a simple to implement, understandable, and widely recognized system is required process for describing the quality of a measurement result, i.e. evaluating and expressing its uncertainty [1].

Error and Its Types

The disparity between the true and observed values is defined as an error. Every measurement has an unknown and unknowable inaccuracy. "Measurement uncertainty" is the term for this unknown inaccuracy. Systematic (or bias) mistakes and random (or precision) errors are the two types of errors [1].

1. Systematic errors (also called *Bias errors*): These are errors that can be made again and over again. The average of measured values minus the true value is the systematic error. There is a specific cause for these problems that can be identified and remedied. Systematic mistake is linked to the phrase "accuracy". As a result, a procedure with high accuracy will have fewer systematic errors.
   Sources of systematic errors:
   - Calibration errors are errors that occur as a result of nonlinearity or faults in the calibration procedure.
   - Loading or infiltration faults—The sensor could really modify the thing it is supposed to measure.
   - When a quantity fluctuates in space, but a measurement is performed only at one site, spatial errors occur (e.g. temperature in a room—usually the top of a room is warmer than the bottom).
   - Human errors can occur if a person habitually reads a scale on the low end of the scale.
   - Defective equipment mistakes occur when an instrument's readings are routinely too high or too low as a result of an internal fault or damage.
   - The mean bias error, defined as MBE=systematic error / true value, is a non-dimensional type of bias error [1].
2. Random errors (also called *Precision errors*): A lack of repeatability in the output of the measurement device causes random errors. The scatter in the measured data is the most common evidence of random mistakes.

Background electrical noise, for example, frequently causes tiny random inaccuracies in the measured output. The reading minus the average of readings is the random error of one data point. Random error is a term used to describe the scatter of repeated measurements around a mean value. The magnitude and sign of the inaccuracy varies at random and cannot be predicted precisely. They are usually easy to spot and are always present in a measurement [1].

As a result, bias error causes changes away from the true mean, whereas precision error induces scatter. The random error of the instrument's output is defined by precision. The reading minus the average of readings is the precision error (of one reading). As a result, precision error is the same as random error [1].

Instrument precision is frequently confused with instrument resolution; however, the two are not synonymous. An instrument's resolution may be excellent, yet its precision may be lacking. The ability of an instrument's output or display to indicate variations in quantity is referred to as resolution [1].

The smallest change or increment in the measured amount that the instrument can detect is formally defined as resolution. The number of digits displayed on the output of digital equipment is frequently connected with resolution. Other defects include zero error, linearity error, sensitivity error, resolution error, hysteresis error, drift error, and instrument repeatability error, in addition to systematic and random mistakes [1].

Calibration and Its Types

Calibration is defined as an operation that, in a first step, establishes a relationship between quantity values provided by measurement standards and corresponding indications with associated measurement uncertainties and, in a second step, uses this information to establish a relationship for obtaining a measurement result from an indication under specified conditions [1].

Types of Calibration

Calibration is broadly divided into two types—static calibration and dynamic calibration.

- *Static calibration*—When time isn't a factor in the measurement, this method is used. As shown in the diagram, some output (voltage, current, etc.) is shown as it varies with a known reference input [1].

  A calibration curve fit is made through the points using several data points at known input values. Any function (linear, parabolic, exponential, etc.) can be used to fit the curve, but most instruments are designed to have linear behaviour [1].

- *Dynamic calibration*—When time is relevant to the measurement, it is essential. The time response of a system or instrument is often determined by abruptly increasing or reducing the input and then recording the output's time response [1].

  Time reaction can be divided into three categories:

  - *Zero order*—The output increases instantly with the input in a zero order (ideal) system. At time t0, the input is abruptly raised. Although no genuine system has a perfect, optimal reaction, this is the goal that instrument designers strive towards.
  - *First order*—A first-order system climbs smoothly to its final value, as depicted, for the identical instantaneous increase of the input at time t0.
  - *Second order*—A second-order system rises to its final value after some time lag for the identical instantaneous increase of the input at time t0, but there may or may not be overshoot, depending on how much damping is available in the measurement system.

    There are generally three possibilities:

    Underdamped—When damping is insufficient, the signal rapidly overshoots and oscillates multiple times before settling to the final reading.
    Overdamped—When the signal is overdamped, it does not overshoot at all, but it takes a long time to settle to the final reading.
    Optimally damped—With just the appropriate amount of damping, the signal overshoots slightly (approximately 5%), then immediately settles to the final reading with very minor amplitude fluctuations [1].

Definition of Uncertainty

Uncertainty is described as "a characteristic associated with a measurement's outcome that characterises the dispersion of the values that may reasonably be attributed to the measurand". The accuracy or uncertainty of all measurement instruments is specified. No instrument can accurately determine the parameter's true value. As a result, the measuring instrument's precision or uncertainty leads to measurement fluctuation. Total variation owing to SWIPE factors (Standard, Work Piece, Instrument, Person & Procedure, and Environment) is sometimes referred to as "uncertainty in measurement", which measures the measurement data's dependability. The less uncertainty there is, the more reliable the data is [2].

Difference Between Error and Uncertainty

The discrepancy between an individual result and the true value of the measurand is termed as error. As a result, while mistake is a single value, uncertainty is a range. Uncertainty is a numerical indicator of the result's quality. Uncertainty is connected with any parameter that cannot be expressed with certainty. It can be found in every quantitative measurement. It is the range of values around the estimated value where the true value of the measured parameter should be found [2].

Sources of Uncertainty

In general, uncertainty can arise from various sources listed in Table 2.1.

There are three main sources of major uncertainty [2] (Fig. 2.1):

- the reference used (measurement standard)
- the unit under calibration (UUC) (measurement result) itself
- the calibration process used (under specified conditions)

Importance of Uncertainty

- It is a numerical indicator of the result's quality.
- Uncertainty information can often save time by avoiding unnecessary analysis repetition.
- There is room for improvement if all components of uncertainty are taken into account.
- It gives useful information regarding the result's quality and consistency.
- Avoids making rash or unsafe decisions based solely on measurements [2].

Components of Uncertainty

It may be required to treat each source of uncertainty independently in order to derive the contribution from that source when calculating the overall uncertainty. An

Table 2.1 Sources of uncertainty

Sampling (stability, contamination, etc.)
Storage conditions
Sample preparation (weighing, sub-sampling, extraction, etc.)
Instrument effects (analytical balance, etc.)
Reagent purity
Assumed stoichiometry
Computational effects
Environmental conditions (dust, humidity, temperature)
Calibration effect (linearity of calibration, weighing, temperature, etc.)
Blank correction, sample effect, operator effect, random effects
Analyst effect (minor variations in applying method, lack of knowledge)

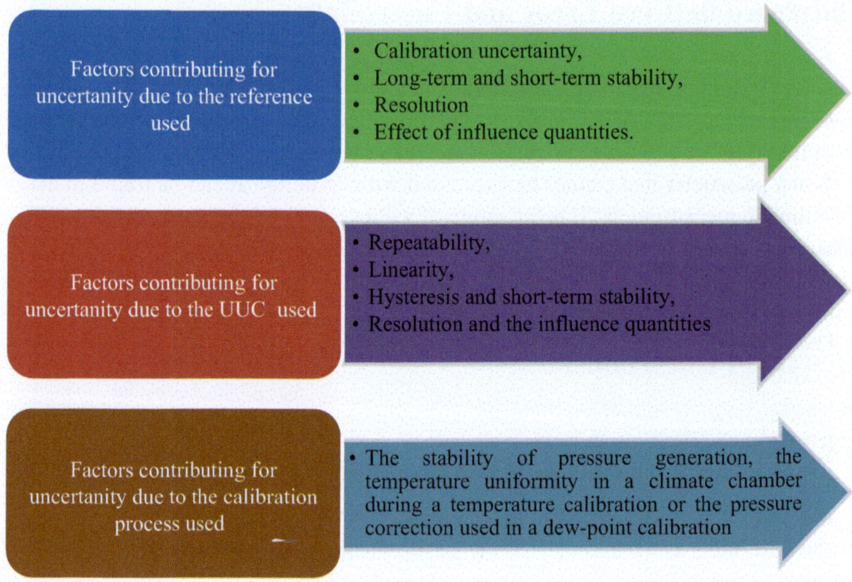

Fig. 2.1 Main sources of major uncertainty

uncertainty component is the name given to each of the individual contributors to uncertainty [3].

- *Standard Uncertainty*—An uncertainty component is known as a standard uncertainty when expressed as a standard deviation. If there is a correlation between any of the components, the covariance must be calculated to account for it.
- *Combined Standard Uncertainty*—It is common to be able to assess the combined influence of multiple factors. This may lower the overall work required, and there may be no need to account for the correlation if components whose contributions are evaluated together are correlated. The total uncertainty, represented by uc (y), for a measurement result y is an estimated standard deviation equal to the positive square root of the total variance derived by combining all the uncertainty components, however evaluated, using the law of propagation of uncertainty.
- *Expanded uncertainty*—An enlarged uncertainty [B.15] U should be employed for most analytical chemistry uses. The enlarged uncertainty provides a range of values within which the measurand's value is believed to lie with greater certainty. The combined standard uncertainty, uc (y), is multiplied by a coverage factor k to get U. The factor k is chosen based on the desired level of confidence. k is 2 for an approximate level of confidence of 95% [3].

Calculation of Uncertainty

Various approaches for error analysis of experimental data exist, including common sense basis, uncertainty analysis, statistical analysis, probability error analysis, limiting error analysis, and so on. The process of discovering and classifying errors is known as uncertainty analysis [4].

Estimating measurement uncertainty necessitates a detailed understanding of the measurement method and its causes of variance, as well as the accuracy and precision of the measurements taken and the integrity of the people who participated in the measurements and calculations [4].

The uncertainty in a measurement's result usually consists of numerous components that can be classified into two groups based on how their numerical value is assessed:

- *Category/Type A*—Those that are examined using statistical methods and are based on any appropriate statistical method, such as average, standard deviation, and so on.
- *Category/Type B*—Those that are evaluated through various methods and are based on scientific judgement based on all available information, such as prior measurement data, manufacturer's specifications, data provided in calibration certificates, and so on [4].

When numerous independent observations for one of the input quantities have been taken under the same measurement conditions, the Type A evaluation of standard uncertainty can be employed. There will be an observable scatter or spread in the values acquired if the measuring method is done with adequate resolution. The experimental standard deviation of the mean, as determined by an averaging process or an appropriate regression analysis, is the standard uncertainty in this case [4].

The process of evaluating standard uncertainty through means other than statistical analysis of a sequence of observations is known as Type B evaluation of standard uncertainty. Standard uncertainty is typically calculated using scientific judgement based on all available data, previous measurement data, experience with or general knowledge of the behaviour and properties of relevant materials and instruments, manufacturer's specifications, data provided in calibration and other certificates, or uncertainties assigned to reference data [4].

The result of uncertainty is expressed as a percentage or in units of measured quantity. Unless otherwise stated, it is always reported at a 95% confidence level [4].

Procedure for Uncertainty Evaluation

- Select the appropriate process after defining and specifying the measurand.
- Determine the sources of uncertainty and the elements that will influence the outcome.
- Identify and quantify the sources of uncertainty.
- Determine the standard level of uncertainty. Type A and Type B evaluations are available.
- Assess the combined level of uncertainty.

- Determine the degree of freedom that is effective.
- Determine the extent of the ambiguity [4].

Methods of Uncertainty Analysis

Type A Evaluations

- Type A evaluation is used to determine uncertainty by taking repeated measurements and calculating the statistical distribution of the data.
- This method is most effective when dealing with random contributions. Measurements with systematic departures from a known right value provide an error value that must be adjusted.
- However, the effect of multiple systematic uncertainties interacts with random uncertainties in such a way that their effect may be identified statistically when analysing the resulting measurement. For example, a systematic temperature offset might create an increase in the measurement result's random thermal noise.
- The standard deviation of repeat measurements, which is approximated by for n measurements with results q_k and average value q, is used in Type A evaluation [5].

$$s(q_k) = \sqrt{\frac{1}{(n-1)} \sum_{k=1}^{n} (q_k - \overline{q})^2}$$

- A single measurement q_k standard uncertainty contribution u_i is given by:

$$u_i = s(q_k)$$

- If n measurements are averaged together, this becomes:

$$u_i = s(\overline{q}) = \frac{s(q_k)}{\sqrt{n}}$$

Type B Evaluations

A Type B analysis is utilized in circumstances where Type A evaluation is unavailable or unfeasible, and to cover contributions not covered in the Type A study [5].

Determine the potential contributions to the total uncertainty in the measurement. Calculate the level of uncertainty for each contribution [5].

- Type A assessment.
- Datasheet from the manufacturer.
- Make an educated guess at a limit value.

Note: Contribution must be expressed in terms of change in the measured quantity, not in terms of influence.

- Choose an expected statistical distribution for each contribution and calculate its standard uncertainty.
- Calculate the extended uncertainty by combining the u_i of that result [5].

There are a number of common distributions for uncertainty contributions:
Normal distribution:

$$u_i = \frac{U_i}{k}$$

Where, *Ui* is the contribution's extended uncertainty, and k is the coverage factor (k = 2 for 95% confidence).

Rectangular distribution: The measurement result has an equal chance of being anywhere between—a_i and a_i.

$$u_i = \frac{a_i}{\sqrt{3}}$$

U-shaped distribution: The likelihood of a measurement result being above or below the median is larger than being at the median [5].

$$u_i = \frac{a_i}{\sqrt{2}}$$

Triangular distribution: Non-normal distribution having a linear decrease in size from maximum to zero [5].

$$u_i = \frac{a_i}{\sqrt{6}}$$

Once the standard uncertainties for all components, including any Type A analyses, have been calculated, the root sum of squares method is used to combine them into a total standard uncertainty (the combined standard uncertainty, uc) for the resultant measurement quantity:

Table 2.2 Uncertainty distribution

Distribution	Factor	Applied for
Normal	1 or 2	CRM [certified reference material]
Rectangular	√3	Balance, temperature
Triangular	√6	Glasswares
U-shaped	√2	–

$$u_c = \sqrt{\sum_{i=1}^{N} u_i^2}$$

Where N is the number of standard uncertainty components. A normal distribution is assumed for the combined standard uncertainty [5] (Table 2.2).

Methods of Uncertainty Analysis

Design-Stage Uncertainty Analysis

It refers to a preliminary examination conducted before to the measurement. It is useful for determining the instrument to use, the measuring procedures to use, and estimating the amount of uncertainty expected to exist in the measurement results [5].

Zero Order Uncertainty

μ0 denotes an estimate of the predicted uncertainty resulting from data reading (interpretation error or quantization error). This fault is expected to be less than the instrumentation error [5].

Arbitrary rule—set μ0 equal to ½ instrument resolution with 95% probability.

μ0 = ±½ resolution (95%).

At 95%, only 1 in 20 measurements would fall outside the interval defined by μ0.

Sequential Perturbation

Using finite difference, this is a numerical method for estimating the spread of uncertainty. It takes the place of the direct method, which necessitates the solution of time-consuming partial differential equations. The method is simple and approximates the derivatives using a finite-difference method [5].

Conclusion

Uncertainty analysis is the process of finding and qualifying errors. Two essential aspects that contribute to certainty estimation are accuracy and precision. To avoid concerns of uncertainty, the idea of uncertainty must be used in the validation of analytical methods. However, there isn't a well-established process for estimating calibration uncertainty. In the future, investigating uncertainty to achieve precision and reliability will be vital and crucial aspects in any process.

References

1. John R (1997) Taylor, an introduction to error analysis: the study of uncertainties in physical measurements, 2nd edn. University Science Books
2. Bevington PR, Keith Robinson D (1992) Data reduction and error analysis for the physical sciences, 2nd edn. WCB/McGraw-Hill
3. Joint Committee for Guides in Metrology (2008) Evaluation of measurement data—guide to the expression of uncertainty in measurement. JCGM 100:1–16. GUM 1995 with minor corrections
4. Duvernoy J (2015) Guidance on the computation of calibration uncertainties. Instruments and Observing Methods 119:1–30
5. Ellison SLR (2000) Quantifying uncertainty in analytical measurement, 2nd edn. EURACHEM/CITAC Guide CG 4, QUAM

Chapter 3
Calibration Methods and Procedures

Abstract A set of procedures known as calibration establishes the relationship between the values reported by a measuring device or measuring system and the standard or known values obtained from the standard under predetermined circumstances. It is a crucial part of every business or organization. Calibration guarantees that equipment is accurate, efficient, and reproducible, as well as providing fitness and maintenance for equipment, instruments, and related services. This chapter covers the fundamental procedures and conditions of the calibration process for any instrument. The key variables to monitor during calibration are temperature, pressure, humidity, and flow rate. Both in-lab and outdoor calibration are options. Laboratory calibration is the most accurate method for calibrating measurement equipment. In comparison to field calibration, laboratory calibration has less uncertainty. Environmental effects are insignificant, and there are significantly fewer factors that can affect calibration. As long as the unit being calibrated and the working standard have sufficient stability times, field calibration is a quick and simple method to examine measuring equipment without having to remove it from the process or process area. Because measurements can affect the final product's quality or safety, calibration is especially important for process-critical devices. The frequency, length of time between calibrations, and type of instrument being used all have an impact on the calibration procedure.

Keywords Calibration · Laboratory · Process · Equipment · Sources · Environment

Introduction

Measurement device calibration is not just a pleasant to have; it is also necessary. To ensure that they generate accurate results, all measuring devices, whether used in production facilities or laboratories, must be calibrated on a regular basis. The process of calibrating an instrument involves making necessary adjustments to ensure that its output perfectly matches its input throughout a predetermined range.

A set of procedures known as calibration establishes the link between the values reported by a measuring device or measuring system and the standard or known values obtained from the standard under predetermined circumstances. An

S. S. Shukla et al., *Pharmaceutical Calibration, Validation and Qualification: A Comprehensive Approach*, https://doi.org/10.1007/978-981-19-9002-1_3

instrument is exposed to known input values while the corresponding output signal levels are being measured in order to make sure that the output precisely matches the input over the course of a continuous range.

To establish recognized input circumstances and measure output signals, trusted standards must be used [1].

Calibration Process

Use a reference standard with established values to encompass the entire range of interest (calibrators). A mathematical function relationship between observed responses and known values of reference standards is called a calibration curve [1].

Steps for Calibration Determination

- Calibrator values
- Model and weight in calibration curve fitting
- Number of replicates at each calibrator level
- Stored calibration curve use

Selection of Calibration Method

The most accurate method for calibrating measurement equipment is laboratory calibration, which is also used in the field. In comparison to field calibration, laboratory calibration has less uncertainty. Environmental influences are insignificant, and there are significantly fewer factors that can affect calibration.

As long as the UUC and working standard have sufficient stability periods, field calibration is an efficient way to test measuring equipment without having to remove it from the process or process area.

The selection of calibration method is listed in Table 3.1 [1].

Procedure of Calibration

The practise of comparing a device under test (DUT) with a reference standard with a known value is known as calibration. Calibration is the process of figuring out an instrument's accuracy in its most basic form. The reference standard, frequently referred to as a "calibrator", is normally traceable to a national or international standard maintained by a national metrology body. It is more accurate than the

Table 3.1 Selection of calibration method

Calibration scheme	Applicable units
Regular calibration	Units in daily service When traceability and uncertainty is important
Calibration before use	Units used infrequently and are not active For new units For repaired units
Calibration if needed	Units used in other service than production, inspection, repairing, and testing of products and can be taken into such service when needed
Inspection before use	Units not calibrated but should be inspected or adjusted before use
No calibration	Units which do not need calibration

equipment being calibrated. The accuracy of temperature, pressure, and level monitoring equipment must be regularly checked because it is essential to pharmaceutical cGMP and quality assurance. Instrument calibration, which is the process of comparing a device's measurements to a standard, can be used to do this. A device that has been validated as accurate and that may be linked to regional or global standards is referred to as a standard. It is possible to alter the test equipment to correct any uncertainty or "drift" from the standard that is noted. For process-critical equipment, calibration is extremely important since measurements might affect the quality or safety of the final product [1].

Pressure Calibration

With gas and hydraulic pressure being the most typical measurements, pressure calibration is an essential task in a range of industries where measurement equipment is required to monitor process performance and safety [1].

Nowadays, many companies have quality standards certifications like ISO9001. Since many industrial processes rely on pressure measurement, pressure calibration is an essential component of a company's quality assurance programme [1]. There are numerous quality procedures that must be followed in order to maintain quality standards.

High-accuracy pressure sensors, pressure gauges, and a variety of pressure balances and calibrators are used to calibrate pressure [1].

Pressure gauges are used in a wide range of commercial contexts, particularly in the pharmaceutical sector. They are used by both manufacturers and laboratories for a wide range of tasks, including as process pressure control, cleaning operations, filters, pumps, monitoring filling levels, and spotting leaks [1].

Instruments for measuring pressure are also essential for safety and quality control, thus they must be dependable, accurate, and precise. Regular instrument calibration benefits pharmaceutical companies because it enables them to maintain the highest level of quality and safety in their manufacturing processes [1].

A few examples of pressure instruments that are regularly calibrated are:

- Digital pressure gauges
- Digital indicators
- Transducers
- Transmitters
- Analogue pressure gauges
- Barometers
- Test gauges

Temperature Calibration

All procedures where temperature readings are important conduct temperature calibration in a controlled environment [1].

Since even slight temperature variations can affect the quality and safety of a product, temperature sensors are crucial for pharmaceutical quality assurance. All temperature-sensing devices, including thermocouple probes, infrared thermometers, RTDs, thermal imaging cameras, and other devices, are delicate and prone to damage [1].

The accuracy and dependability of these devices must be regularly verified, and the results of these testing must be well documented to assure traceability. Frequent calibration supports these procedures as well as the prompt identification of any measurement drift or variation brought on by corrosion, rough handling, or accidental damage. Temperature calibration frequently uses resistors, thermocouples, or platinum resistance thermometers (PRTs), also referred to as resistance temperature devices (RTDs) [1].

It is crucial to understand that utilizing an RTD or thermocouple indicator to measure a temperature sensor's output and comparing the results to the in-line field indication does not constitute temperature calibration. Only by comparing the probe being tested to a recognized reference in a stable temperature environment can a temperature calibration be performed [1].

A few examples of equipment that require temperature calibration on a periodic basis are:

- Data Acquisition Systems
- Spring-Type or Bimetal Thermometers/Thermocouples
- Dial Thermometers
- Chambers/Furnaces
- Infrared Meters
- PRTs and Thermistors
- Thermal Imaging Cameras

- Resistance-Temperature Detectors (RTDs)
- Infrared Radiation (IR) Detectors

A variety of procedures and testing standards can be used to calibrate temperature sensors, including:

- Fixed-point cells and intrinsic standards are typically exclusively employed by calibration laboratories due to the prohibitive cost of internal calibration. They offer the most accurate temperature sensor calibration available.
- Liquid bath calibrators: These high-maintenance calibrators are often installed in stationary locations such as calibration labs and are utilized for extremely small sensors or devices with unusual shapes.
- Dry-block probe calibrators are more portable, require less upkeep, and reach the desired temperature more quickly than liquid bath calibrators, but they are less accurate and stable.
- The most affordable and portable calibrators for in-house testing are electronic ones, but they lack a temperature source to determine the sensor's integrity [1].
- Pyrometers, thermal cameras, and other noncontact temperature sensing tools are calibrated using infrared calibrators.

Flow Calibration

The linear, nonlinear, mass, or volumetric flow rates of a liquid or gas are measured using a flowmeter, also referred to as a flow sensor. Control and instrumentation engineers measure the flow rate, which is the rate at which a process fluid passes through pipelines, orifices, or vessels at a specific moment, in order to monitor and control the quickness and effectiveness of industrial flow processes and devices [1].

For flow equipment to help enhance productivity, profitability, and regulatory compliance, calibration is necessary.

These flowmeters that confirm the quality and quantity of the product or feedstock, the amount of fuel or energy, or that are used in a critical process require routine flow calibration services to ensure that measurements are accurate and enable operations to run without incident and on schedule [1].

Level measuring is a method of determining the location or level of a surface inside vessels like tanks and reactors for precise and reliable inventory control.

Level measurement is just as important in tank gauging systems as temperature and pressure measurement. These systems offer advantages in terms of operations and safety, as well as being useful for estimating net quantities and inventory. Depending on the application, level measurement may be done using continuous level transmitters or point level devices [1].

Instrument calibration can help ensure that level measurement findings are precise, repeatable, and traceable to industry standards, just like other process-critical measurements.

The four main types of flowmeters that frequently require calibration are:

- Thermal Mass Flowmeters
- Laminar Flowmeters
- Rotometers—Gas and Air
- Turbine Meters

Pipette Calibration

For accurate and precise pipetting results in laboratories that often use this measurement equipment, pipette calibration is essential. All pipettes used in laboratories, including single-channel, multichannel manual, and electronic pipettes, must be calibrated according to certain procedures. Pipette calibration's main objective is to ensure that dispensing is carried out with the desired level of precision [2].

Electrical Calibration

Electrical calibration is the process of assessing the performance of any instrument that measures or tests electrical characteristics like voltage, current, resistance, inductance, capacitance, time, and frequency. Electrical calibration demands the employment of precise tools or calibrators to evaluate the performance of crucial characteristics of additional devices known as units under test (UUTs) [2].

Instruments that are often sent for electrical calibration are:

- Data loggers.
- Electrical meters.
- Multi-meters.
- Oscilloscopes.
- Frequency counters.
- Insulation testers.
- Loop testers.

Mechanical Calibration

Mechanical instruments require mechanical calibration because they are prone to drift from repeated use, mechanical damage, and exposure to changing air conditions [2].

Mechanical calibration takes place in a temperature-controlled environment, calibrating mass, force, size, angle, volume, flatness, torque, and vibration [2].

Among the tools that are most frequently calibrated mechanically is:

- accelerometers
- scales/balances
- load cells and force gauges
- micrometers, verniers, height gauges
- torque wrenches and screwdrivers
- weight and mass sets

Key Requirements for Calibration Process

1. special attention to the design of equipment/instrument to be calibrated
2. check the tolerance
3. maintaining accuracy ratio
4. adhering to standards

Equipment can be shipped to a calibration lab for calibration or calibrated on-site. Depending on the frequency, quantity of devices being inspected, and type of equipment being used, it could be simpler to use calibration services to ensure measurement instruments are accurate and their results are traceable [2].

Accredited labs offer on-site calibration, which can reduce costs associated with obtaining testing supplies and instructing staff members in calibration and testing procedures. This can soon add up to a large cost because the majority of pharmaceutical companies use a range of temperature, pressure, and level sensing equipment [2].

Additionally, time can be saved. In-house calibration can cause a lot of downtime, but calibration services labs have the expertise, tools, and training to minimize the amount of time that the equipment is out of commission [2].

Calibration Frequency

Every measurement instrument must be calibrated on a regular basis based on a number of factors, including quality standards, damage risk, usage frequency, and drift rate. Develop a calibration schedule based on the types of equipment used and the unique process requirements by consulting a certified lab [3]. The instrument manufacturer's suggestion, the frequency and severity of predicted use, the impact of the environment, the maximum allowable deviation of the measurand, and the necessary measurement uncertainty are all taken into consideration when deciding on the calibration interval.

Sources of Calibration

It is believed that calibration is a sacred measurement that should only come from reliable sources. When selecting a calibration laboratory, keep the sources listed below in mind. Priority should be given to a calibration laboratory that has received ISO/IEC 17025 accreditation. The website of the organization responsible for accrediting the laboratories in question contains a list of these facilities.

The website address of a particular country's accrediting body can be found on the ILAC or APLAC websites because the majority of nations with a laboratory accreditation system are members of either the International Laboratory Accreditation Cooperation (ILAC) or the Asia Pacific Laboratory Accreditation Cooperation (APLAC). Check the accreditation's scope before selecting a lab to see if it can calibrate a certain measuring gadget.

The following laboratory in line is one that has submitted an application for accreditation or has obtained certification in accordance with the ISO 9001:2000 standard on "Requirements for a quality management system" [3].

- If neither of the aforementioned categories has any calibration labs, the lab should be selected by evaluating the following factors:
- The traceability of the standard.
- Test reliability ratio is accurate. This is the proportion of the accuracy of the calibration standard to the stated accuracy of the calibrated measuring apparatus. This ratio must be at least three to one.
- Extensive details regarding the real worth of the study. In order for the user of the measuring equipment to understand how much variation there is between the exhibited reading and the corresponding standard value, the variation between the shown reading and the corresponding standard value should be supplied on the report.
- Measurement uncertainty.
- The accessibility of qualified labour.
- A good name in the market [3].

Calibration Laboratory

Calibration laboratories that use the AS 17025 or ISO 17025 quality management system must adhere to all NIST traceability and quality standards in order to receive accreditation. It works in a similar way to how ISO 9000 for manufacturing organizations makes sure that calibration services are performed correctly.

Services and traceability must be independently verified by businesses. It is possible for unaccredited labs to provide good service and traceability.

Using authorized service providers can save time, money, and effort by ensuring that these labs are following the correct techniques and accurately recording operations. A reputable lab can provide the appropriate calibration certifications for each piece of equipment in order to avoid issues during regulatory inspections.

FDA regulations and pharmaceutical cGMP must be followed, and failing to do so can lead to penalties, fines, and even product recalls. Using any measurement equipment requires accurate results. The use of accurate calibration is the only way to define accuracy and guarantee that measurements are recorded in accordance with guidelines [4].

Calibration Based on Types of Instrument

Linear Instruments

The most fundamental calibration process for an analogue, linear device is known as the zero-and-span approach. This is how it is done:

- Give the instrument the lower range value stimulation and wait for it to stabilize.
- Adjust the "zero" setting until the instrument registers correctly.
- Give the instrument the upper-range value stimulation and wait for it to stabilize.
- Adjust the "spread" setting until the instrument registers correctly.
- If necessary, repeat steps 1 through 4 to attain good accuracy at both ends of the range.

By testing the instrument's response multiple times between the lower and upper range values, this crude method is made better. The "five-point calibration", which entails evaluating the instrument at 0% (LRV), 25% (50%), 75% (URV), and 100% (URV) of range, is a famous example of this [2].

Another approach to this problem is to check at 10%, 25%, 55%, 75%, and 90% while continuing to make zero and span modifications at 0 and 100%. Checking the instrument's behaviour at five calibration points while increasing and decreasing is an additional enhancement above the conventional five-point test. This kind of testing is frequently referred to as calibrations from the top down. This test will establish whether the instrument has any hysteresis (a lack of responsiveness to a change in direction) [2].

The linearity of several analogue instruments can be changed. These linearity adjustments can be rather sensitive, and too eager hands have a tendency to overcorrect them.

Only alter an instrument's linearity if its entire range cannot be used to attain the needed precision. Alternatively, disregard linearity and modify the zero and span parameters to "divide" the error between the scale's greatest and lowest points [2].

Trimming differs differently from calibrating a "smart" digital transmitter in terms of approach. The "low" and "high" trim procedures of a digital instrument are typically non-interactive, in contrast to the zero and span adjustments on an analogue instrument [2]. The four standard techniques for trimming a "smart" instrument's sensor are as follows:

- Before utilizing the "low" sensor trim instruction, stimulate the instrument with lower range values.
- Wait for it to stabilize.
- To offer the instrument upper-range value stimulation, select the "high" sensor trim option.
- Wait for it to stabilize.

Trimming a "smart" instrument's output (Digital-to-Analog Converter, or DAC) follows the same six broad steps:

- Run the output trim test at the "low" level.
- After the output signal has stabilized, measure it with a precision milliammeter and note the result.
- When the instrument asks you, enter this measured current value.
- For the output trim test, select "high".
- When the output signal has stabilized, use a precision milliammeter to measure it and record the value [2].
- Enter the most recent value you measured when the instrument asks.

A smart transmitter's lower and higher range values can be changed after both the input and output (ADC and DAC) have been decreased (i.e. calibrated against known-to-be-accurate standard references). The transmitter can be ranged and ranged again as needed after the trim procedures are finished. Re-trimming a smart transmitter is mostly done to maintain accuracy over extended periods of time when the sensor and/or converter circuits have strayed outside of permitted bounds. In contrast, the recalibration required by analogue transmitter technology for re-ranging [2].

Nonlinear Instruments

It is far more difficult to calibrate nonlinear equipment than linear instrumentation.

Two adjustments (zero and span) are no longer adequate to describe a curve because more than two points are needed. Nonlinear devices include expanded-scale electrical metres, square root characterizers, and position-characterized control valves [5].

When calibrating a nonlinear instrument, it is important to record all changes made so that, if necessary, the instrument can be "re-set" to its initial state [5].

Discrete Instruments

The definition of "discrete" is unique or distinct. In engineering, a "discrete" variable or measurement is referred to as a true-or-false condition. Therefore, a discrete sensor can only provide information on the measured variable's position in relation to a preset set point [5].

Process switches, for instance, are discrete devices that activate and deactivate at particular values. An analogy for a discrete instrument would be a pressure switch that activates an air compressor when the pressure drops below 85 PSI [5].

Like continuous instruments, discrete instruments need to be calibrated on a regular basis. On the majority of discrete devices, the set-point or trip-point is the only calibration adjustment possible. Some process switches include two adjustments: a set-point adjustment and a dead band adjustment [5].

When calibrating a discrete instrument, verify the precision of the set-point in the direction of stimulus change. This entails making sure the switch changes states when the pressure falls below 85 PSI rather than rising over 85 PSI in the instance of our air pressure switch.

If the dead band weren't there, it wouldn't matter which direction the applied pressure changed during the calibration test. Contrarily, a discrete instrument will always have a dead band, regardless of whether it is adjustable [5].

One method to effectively calibrate a discrete instrument without a lot of trial-and-error attempts is to set the stimulus to the desired value (e.g. 85 PSI for our hypothetical low-pressure switch), and then move the set-point adjustment in the opposite direction as the intended direction of the stimulus (in this case, increasing the set-point value until the switch changes states). This technique is based on the finding that the majority of comparison systems cannot distinguish between a rising process variable and a falling set point (or vice versa).

Uncertainty Estimation

There are three main causes of uncertainty:

- the reference utilized
- the unit under calibration (UUC)
- the calibration procedure used

Here is a description of some of the origins of uncertainty. The reference uncertainty is made up of the calibration uncertainty, long- and short-term stability, resolution, and the impact of influencing quantities. The UUC's uncertainty is made up of repeatability, linearity, hysteresis and short-term stability, resolution, and influence quantities. The calibration process itself may be to blame for the inaccuracy of the height correction used in a pressure calibration, the temperature homogeneity in a climate chamber during a temperature calibration, or the pressure correction used in a dew point calibration [6, 7].

Conclusion

All measurement tools are inaccurate, and calibration determines how inaccurate they are. Calibration has become a need in everyday life for achieving and obtaining the highest performance in product manufacturing, production, and output. A calibration curve is a mathematical function relationship between observed responses and known values of reference standards. Equipment should be calibrated before use, after use, and whenever necessary; however, if calibration is not required, some equipment may be used without calibration. In its most basic form, calibration is the process of determining an instrument's accuracy. Special attention to the design of the equipment/instrument to be calibrated, checking the tolerance, maintaining the accuracy ratio, and adhering to standards are the primary requirements for the calibration process. So that difficulties do not arise during regulatory inspections, an authorized lab AS 17025 or ISO 17025 can give the proper calibration certifications for each piece of equipment. Failure to produce valid paperwork is a violation of FDA laws and pharmaceutical cGMP, and failure to do so can result in penalties, fines, and even product recalls. The calibration procedure varies depending on whether the instrument is linear, nonlinear, or discrete. Uncertainty must also be measured, which comes from three basic sources. This will give the equipment a fresher look while also extending its life and improving its performance.

References

1. International Organization for Standardization (1993) International vocabulary of basic and general terms in metrology, 2nd ed
2. International organization for Standardization. Measurement management systems: requirements for measurement processes and measuring equipment (ISO 10012:2003)
3. American National Standards Institute/National Conference of Standards Laboratories. Calibration laboratories and measuring and test equipment: general requirements (ANSI/NCSL Z540–1-1994)
4. International Organization of Legal Metrology. Guidelines for the determination of recalibration intervals of measuring equipment used in testing laboratories (document D10, currently being revised). See the website of Deutsches Akkreditierungssytem Prufwesen GmbH at http://www.dap.de/95doc/DAP-TM-08.pdf
5. https://instrumentationtools.com/instruments-calibration-procedures. Accessed 26 Jan 2022
6. Expression of the uncertainty of measurement in Calibration. EA-4/02.
7. International vocabulary of basic and general terms in metrology/ISO 1993 10. Guide to meteorological instruments and methods of observation. WMO No. World Meteorological

Chapter 4
Calibration Software Support

Abstract A complete calibration system that includes tightly integrated hardware and software components should have calibration management software at its core. The examination of calibration records using calibration management software is quicker, simpler, and more precise. It also makes it possible to spot historical trends. Software-based calibration management improves planning and decision-making. All of these calibration management processes—planning and decision-making, organization, execution, documentation, and analysis—benefit from the use of calibration software. Two devices are compared during calibration, with one of the devices having a known accuracy. Outside of a calibration laboratory, software is used to calibrate instruments in the field or in the lab. The main objective is to control the climate in order to limit the effects of temperature, air pressure, and humidity. The main procedures needed for calibrating any equipment include scheduling, resource management, cost analysis, and storing all data and documentation to the satisfaction of an auditor. For a number of reasons, including ensuring that production output fulfils set quality parameters and complying with regulatory requirements, calibration and maintenance of process site instruments are essential. The two primary types of calibration software are software from calibrator manufacturers, which can be closely coupled with the manufacturer's hardware, and software from independent software firms, which does not include hardware solutions.

Keywords Calibration · Software · Hardware · Maintenance · Management

Introduction

Two devices are compared during calibration, with one of the devices having a known accuracy. For the accuracy to be of any real significance, it must be traceable to a nationally or globally accepted reference instrument that has been shown to provide the smallest possible difference from the absolute actual value.

The plant used to be shut down once a year for instrument calibration. This tactic's drawback was that every instrument, to some extent, drifted from specification, and drift may happen at any time over the 12-month working period.

S. S. Shukla et al., *Pharmaceutical Calibration, Validation and Qualification: A Comprehensive Approach*, https://doi.org/10.1007/978-981-19-9002-1_4

The majority of testing and calibration laboratories operate their testing and calibration operations using a laboratory information management system today [1] and are moving towards more sophisticated management [1]. However, there are many areas where data connectivity and retrieval might be improved, including quality control, sample flow management, and electronic original documents, to name a few.

Difference Between Calibration and Maintenance

For a number of reasons, including ensuring that production output fulfils set quality parameters and complying with regulatory requirements, calibration, and maintenance of process site instruments are essential. It is also necessary for the creation of a successful traceability system. It is common to think of calibration and maintenance as two different jobs that need to be done separately. To ensure that the calibration results are incorporated into maintenance procedures and that maintenance plans are scheduled based on the calibration results, they should be completed simultaneously for the greatest effectiveness [1].

Calibration Management Software

Software is used to calibrate instruments in the field, i.e. away from a calibration laboratory, or in a lab. The main objective is to control the climate, including the temperature, pressure, and humidity. Modern laboratory calibration tools are built with software interfaces so they can carry out both laboratory and field calibrations. The programme also records the history of setup, calibration, and maintenance. This is updated each time a calibration is carried out [1].

A user-friendly Windows Explorer-like interface with specialized calibration management software is provided to users. The programme manages and stores all instrument and calibration data. This includes communication with smart calibrators, planning and scheduling of calibration tasks, frequency analysis and optimization, report, certificate, and label printing, and simple integration with CMM systems like SAP and Maximo. When calibration data is properly analysed, compliance, efficiency, quality, and safety can all be improved [1].

The general stages for performing calibration on any instrument are as follows:

- scheduling
- resource management
- cost analysis
- storing all the data and documentation to the satisfaction of an auditor

The entire calibration system of a customer is growing more and more depending on calibration management software (CMS). CMS increases the productivity of the calibration process, eliminates errors, and lowers total expenses [2].

Function of Calibration Management Software

- the creation of hands-free automated calibration procedures
- the generation of automatic work schedules for calibration engineers
- the recording and storage of historical calibration results
- the analysis of sensor calibration trends
- audit trials for ISO9001 and other quality standards [3]

Advantage of Calibration Management Software

- The ideal calibration period for any piece of equipment can be determined by analysing the calibration data that has been gathered.
- Assures that the system is audit-ready to satisfy statutory and regulatory requirements.
- Modern software can also point out places where calibration periods could be extended without the device failing.
- Ensures the safety of the plant and has a favourable impact on quality and efficiency.
- Documentation functions are typically found in calibration management software, allowing users to create reports and calibration certificates using information gathered from connected calibration equipment or test devices [3].

The best calibration management software also has features for resource management and efficiency improvement. Using an interval analysis module, users can provide reports on the operation of the device by objectively analysing historical data and drift. By offering quick and reliable information and analysis, this tool helps calibration managers who are in charge of expanding calibration intervals make an informed decision. Similar to this, real-time updates to Key Performance Indicators (KPIs) enable study and comparison of installed base and resource performance across the entire organization or at various levels [3].

By ensuring that calibration failure notifications are issued in a timely and controlled manner, vision deviation management enables users to track down the root cause of deviations, investigate the issue, and find a solution. The audit trail will then demonstrate that a strong deviation management strategy is in place, along with strict rules for corrective and preventative action [3].

Need of Calibration Management Software

Every manufacturing plant uses a system to manage instrument calibration procedures and data. To ensure that they are operating properly and measuring within predetermined tolerances, plant instrumentation devices including temperature sensors, pressure transducers, and weighing instruments must undergo routine calibration [3].

On the other hand, numerous businesses from various industry sectors handle these calibrations in remarkably different ways. These systems have vastly different costs, data quality, efficiency, and accuracy levels, as well as levels of automation [3].

A product that can be used to support and direct calibration management duties is calibration software, with documentation being a key component [3].

There are five main areas here, comprising of planning and decision-making, organization, execution, documentation, and analysis.

- *Planning and decision-making*: List every instrument and measurement tool used in the plant, then classify each one as "critical" or "non-critical". The requisite tolerances and calibration range must then be established. Following that, choices must be made on the timing of each device's calibration. Standard operating procedures (SOPs) for each device must then be created and authorized, after which the best calibration techniques and related tools must be chosen. Last but not least, the business needs to figure out the calibration status of each instrument in the facility [4].
- *Organization*: It comprises imparting knowledge on how to use the chosen tools and follow the approved SOPs to the company's calibration team, which typically consists of maintenance technicians, service engineers, process and quality engineers, and management. To carry out the planned calibration tasks, the resources must next be organized and assigned [4].
- *Execution*: In this phase, you must oversee the calibration duties that were given to you. The personnel doing these duties must adhere to the necessary regulations, including any applicable safety standards, before calibrating the equipment. The calibration is then completed as intended, albeit it is possible that extra instructions are needed after calibration [4].
- *Documentation*: Documentation is a crucial part of the calibration management process. According to ISO 9001:2000 and the FDA, calibration records must be preserved and calibration must be carried out using established, recognized techniques. For documentation and the preservation of calibration results, all calibration records must be signed and approved. Planning, creating, and pasting calibration labels, as well as copying and preserving produced documents, are required for the following calibration tasks [4].

Types of Calibration Software

There are basically two types of calibration software –

- software from calibrator manufacturers that is tightly connected with the hardware of the manufacturer

- software from independent software businesses that does not come with hardware solutions.

Many organizations want to implement fully integrated software-based solutions, but many are experiencing difficulty obtaining the IT infrastructure and support they require. High-end computerized maintenance management systems (CMMS) have been used by some businesses to schedule maintenance work but not to organize calibration resources, create calibration processes, or gather and analyse calibration data [4].

The best software will typically integrate calibration and maintenance management and offer a method to improve maintenance workflow by improving data integrity, planning work activity, automating calibration and maintenance, and identifying and correcting errors. After that, the programme offers a fully electronic system with e-signature support for regulatory compliance. Another advantage is that it might connect to bigger plant-level systems and provide data that is ready for audits and complies with the strictest quality procedures [4].

Calibration software like MET/CAL is one of the links in the calibration chain. It is a good idea to test the software after installation on the user's computer network, even when extensive testing is done at the software developer's facilities. This would guarantee a smooth installation, software compatibility with user hardware, and the ability of the software to deliver precise measurement data to file storage during a calibration verification or adjustment event. The best calibration management software also has the capacity to optimize efficiency and manage resources. An interval analysis module, which conducts a scientific analysis of historical data and drift, allows users to report on device performance. This application provides quick and reliable reports and analysis to help calibration managers who are in charge of expanding calibration intervals make an informed choice [4].

Conclusion

Instruments, equipment, and related services are used by pharmaceutical manufacturing enterprises to manufacture high-quality products with the greatest efficiency. All plant instruments and measurement devices are identified and categorized as "critical" or "non-critical". Measurement equipment, sample detection, operation record, certificate life cycle management, multichannel customer notification, and intelligent early warning are all areas where the calibration management information system excels. It ensures a higher level of refined management, standardization of corporate processes, increased work productivity, ongoing quality management, and better customer service. In the future, software calibration will be required in every business to preserve procedure and obtain the best potential results.

References

1. ISO (2017) General requirements for the competence of testing and calibration laboratories. ISO/IEC 17025:2017
2. Binnebosel M, Trotha KTV, Steinau G, et al (2011) Testing and calibration laboratories implement delicacy management based on LIMS. Modern Scientific Instruments
3. Guidance for Industry, General Principles of Software Validation, Draft Guidance Version 1.1, Section IV.G
4. Software validation for pharmaceutical and medical device manufacturers, section 1.5

Part II
Qualification

Chapter 5
Objectives of Qualification

Abstract The qualification of equipment is an integral part of the quality assurance and GMP processes. It is an important part of the validation procedure. The main function of any equipment after purchase is to ensure consistent quality of final goods as well as regular use. Qualifying, in layman's words, is the process of testing equipment, instruments, services, utilities, and related systems. A collection of actions carried out or completed in order to verify and confirm that the equipment and utilities are fit for their intended purpose and perform correctly is referred to as qualification. Safety, integrity, strength, purity, and quality are used to classify equipment as critical or non-critical. The quality assurance procedure and GMP include equipment qualification. It is a requirement of the validation procedure. After purchase, the primary goal of any piece of equipment is to ensure consistent quality of finished goods and regular use. Qualifying is the testing of equipment, instruments, services, utilities, and associated systems in layman's words. A collection of actions carried out or completed to verify and confirm that the equipment and utilities are fit for their intended purpose and function properly is referred to as qualification. Based on its safety, integrity, strength, purity, and quality, equipment is classified as critical or non-critical. The Factory Acceptance Test is usually performed at the vendor's site during the specification phase of qualification, while the Location Acceptance Test is usually performed at the purchaser's site. The criticality of the equipment and system determines the level of qualification.

Keywords Equipment · Qualification · Quality assurance · Performance · Requalification

Introduction

In the pharmaceutical sector, the use of qualified equipment in accordance with Good Manufacturing Practices and Quality Assurance is required in every department, particularly in the production and packaging of pharmaceuticals, in order to achieve the best medicinal product. Their primary goal of any equipment after its purchase is to ensure consistent quality of final products along with its frequent use. The set of operations undertaken or conducted to demonstrate that the utilities and

S. S. Shukla et al., *Pharmaceutical Calibration, Validation and Qualification: A Comprehensive Approach*, https://doi.org/10.1007/978-981-19-9002-1_5

equipment are suitable for their intended purpose and are functioning properly is known as qualification. The term qualification is defined as "the action of proving and documenting that equipment or ancillary systems are properly installed, work correctly and consistently, and actually lead to the expected results", according to the EU GMP guidelines; Part 2 deals with the Basic Requirements for Active Substances Used as Starting Materials. Qualification is similar to validation and is included in the validation process; however, individual qualification processes do not constitute process validation". Qualification guarantees that a system complies with regulatory requirements, industry standards, and performance expectations. The degree of qualification is mostly determined by the equipment's complexity [1].

Objectives of Qualification

- To ensure precise and accurate measurement of analytical data.
- It is a continuous procedure that leads to the desired result.
- Qualification ensures that a certain instrument/equipment and associated system meets regulatory criteria, industry standards, and expected performance [1].

Benefits of Qualification

- Reduces the chances of a system malfunctioning, as well as the likelihood of an incorrect test.
- A clearer picture of the procedure.
- Ensure that the process runs smoothly.
- Reduces the range of possible outcomes.
- Ensure that product quality is maintained and controlled.
- Reduces the number of defects and makes the process more cost-effective.
- Improves the overall reliability and availability of production.
- Reduces the likelihood of a product recall and failure.
- Ensures that the procedure is both safe and efficient.
- Regulatory audits are saved due to non-compliance [1].

Need/Requirement of Qualification

The need for qualification for any individual piece of equipment is determined by its intended use. For new instruments, equipment, services, utilities, and facilities, qualification is carried out.

Qualification in instrument/equipment/services/utility/facility is required after any modification or breakdown. Changes or extensions to the qualification range, as well as periodic requalification, are also required.

Difference between Validation, Calibration, and Qualification

Validation, calibration, and qualification are all phrases that refer to the same thing when it comes to equipment. These are required for equipment in order to maintain its efficiency and performance [2].

The difference between them is enlisted in Table 5.1.

Table 5.1 Difference between validation, calibration, and qualification

Validation	Calibration	Qualification
Action of proving, in accordance with the principles of good manufacturing practice, that any procedure, process, equipment, material, or activity. Actually leads to the expected results	Calibration is a demonstration that, a particular instrument or device produces results within specified limits by comparisons with those produced by a reference or traceable standard over an appropriate range of measurements	Action of proving and documenting that equipment or ancillary systems are properly, installed, work correctly, and actually lead to the expected results
Validation is a method of quality assurance and an important part of the GMP	In calibration performance of an instrument or device is compared against a reference standard	Qualification is part of validation, but the individual qualification steps alone do not constitute process validation
With the validation, the performance, quality, and other operating parameters of a system are tested to verify that they comply with the requirements	With the calibration, the measurements are compared with an accepted reference measurement, to assure the considered measurements comply with the requirements	Qualification of analytical instrumentation is essential for accurate and precise measurement of analytical data. If the instrumentation is not qualified, ensuring that the results indicated are trustworthy, all other work based upon the use of that instrumentation is suspect

Relationship Between Validation, Calibration, and Qualification

Qualification and validation are both based on the same principles. Qualification, on the other hand, is seen as a component of validation. Qualification is used to describe new and existing equipment and utilities, whereas validation is used to describe a process, technique, or system. Validation is usually preceded by qualification. One of the most important aspects of guaranteeing the legitimacy of Qualification and Validation is calibration [2].

Types of Equipment

In Regulatory Compliance and GMP, qualification is critical. To ensure product validation and qualification, the parameters Safety, Integrity, Strength, Purity, and Quality Assurance are critical. According to GMP, equipment is categorized into two groups based on safety, integrity, strength, purity, and quality [SISPQ] [2].

Critical Equipment

Customized equipment are another name for these. Critical equipment are those that are required for the processing, packing, holding, or support of products, as well as those unit actions that have the ability to affect critical process parameters and product quality.

They have a direct effect on SISPQ, and their frequency is determined by usage and risk assessment in accordance with regulatory guidelines. All qualification documents, such as User requirement specification [URS], Factory Acceptance Test [FAT], Site Acceptance Test [SAT], Design qualification [DQ], Installation Qualification [IQ], Operational Qualification [OQ], and Performance Qualification [PQ], are necessary for important equipment [2].

Non-critical Equipment

Critical equipment are those that are required for the processing, packing, holding, or support of products, as well as those unit operations that have no direct impact on critical process parameters or product quality. They have a direct effect on SISPQ, and their frequency is determined by usage and risk assessment in accordance with regulatory guidelines. Documents linked to system or utility qualification are not required for non-critical equipment, but a record of the installation report must be kept. Non-customized or off-the-shelf equipment are another name for them. The records URS, SAT, FAT, and DQ must be kept up to date according to the requirements [3].

On the basis of the risk factor connected with them, equipment, including instruments, can be divided into three categories. These are as follows:

- Low-risk apparatus [Category A]: Any standard piece of equipment that does not require measurement or calibration. By confirming the data, which is confirmed and documented by observation of operation, such equipment meets the qualifying criteria. Vortex shakers, mixers, sonication baths, evaporators, and magnetic stirrers are examples.
- Medium-risk equipment [Category B] includes any standard equipment that provides measurement or reading while in use. Qualification, calibration, and timely verification are all required for such equipment. pH meter, balance, thermometer, centrifuge, pumps, and so on.
- High-risk equipment [Category C]: This category includes all sophisticated and advanced machinery, such as complex sensitive equipment/instruments and computerized systems. These devices must be fully qualified before being subjected to functional and performance tests. Analytical, chromatographic, and spectroscopic equipment, such as HPLC, GCMS, FTIR, and zeta sizer, are examples [3].

Life Cycle of Equipment and Steps for Qualification of Equipment

Any equipment from manufacturing to final purchase, i.e. from owner to vendor site involves all four phases of qualification. The steps for qualification of equipment includes—

- preparation of protocol
- preparation of schedule
- coordination and training as per schedule
- execution of protocol
- collection of data
- review of data collected
- review and approval of deviation
- compilation of report
- preparation of summary of qualification
- updation of the status label and schedule
- approval of executed report [4]

Phases/Elements of Qualification

The phrase "qualification" refers to new equipment and utilities. In most cases, according to WHO, qualification must be fulfilled before the process can be validated. To put it another way, qualification comes before validation. It is carried out in two stages for devices, equipment, and analytical instruments [5].

Specification Phase

This is the first step in the Qualification process, which includes initial documentation in the form of a User Requirement Specification, followed by Design Qualification [5].

Verification Phase

After the specification phase is completed, the verification phase begins, which includes the installation, operation, and performance study (IQ, OQ, and PQ). For appropriate and optimal performance in any industry, any equipment requires calibration, validation, and qualification. These three should be done on a regular basis. Throughout the life cycle of an item's functioning, it is subjected to documented periodical performance checks. In simple words, the 4Q technique refers to the four main stages of equipment qualification [DQ, IQ, OQ, and PQ] [5].

Phase I

The user specification requirement is the first and most basic document that streamlines the entire qualifying process. These are created by the final user, who outlines their process's expectations and requirements. A policy for equipment and system qualification must be established by the manufacturer. The URS has a significant impact on the overall certification process. The acceptance criteria for URS includes–

- URS approval sheet
- Table of content
- Purpose and scope of URS
- Document history
- Description and overview
- User requirement
- Functional specification

- Operational specification
- Design specification
- Electrical specification
- Software and hardware specification
- Specific requirement
- Documentation requirement
- Scope of work
- Submission of design to user for review and approval [5]

The DQ, or Design Qualification, comes after this URS. Design Qualification is the documented verification that the planned design of the equipment and related system is adequate and suitable for the intended purpose. Two crucial DQ criteria are as follows: When designing any equipment or system, the user requirements must be taken into account at all times [5].

For the specific equipment/system, an approved vendor or suitable provider should be identified and selected in advance [5].

Design qualification (DQ) is a defined set of actions that specify the instrument's functional and operational specifications, as well as vendor selection criteria, depending on the instrument's intended purpose. Design qualification (DQ) is a process that can be carried out by both the manufacturer and the end user of the equipment. The manufacturer is totally responsible for the specification of the design of equipment, its functioning, operation, maintenance, and robustness in order to illustrate and characterize that specific instrument/equipment. For periodic maintenance, training, updates, installation services, and system complexity, the user should coordinate and be in contact with the manufacturer. The DQ is separated into three parts in general: proposal evaluations, risk analysis and test setup, correct setup, and management/maintenance [6–8].

The acceptance criteria for DQ includes—

- details of equipment drawing
- technical specifications of the equipment
- construction material and related details
- utility requirement
- operational specifications
- UPS/emergency power requirement
- computer system and related logical programmes details communication with other system/equipment
- interlock details
- design specification
- electrical, pneumatic, and instrumental drawing
- control description
- compliance details with reference to 21 CFR—Part 11 [6–8]

The traceability matrix, which functions as a link between risks and the requirements accomplished through numerous tests, is the final output that delivers the result of DQ.

The Factory Acceptance Test (FAT)—After acquiring equipment, it is the vendor's responsibility to perform a FAT test before shipping the instrument. The FAT is a document that certifies that the equipment was built/manufactured in accordance with the specifications and functionalities, and that it is ready for delivery. URS, DQ, Purchase specification, Design drawing and engineering diagrams, Safety standards, Software requirement as per CFR 21 Part 11, and trial runs should all be met by the equipment and associated system.

The Site Acceptance Test—This test is typically performed at the purchaser's location to demonstrate that the item performs as expected. Only complicated devices and huge projects should use a FAT in conjunction with SAT. The following are the minimum approval criteria: major component verification according to FAT and requirements:

- list of packaging materials
- manufacturer's machine identification number/code
- physical condition of primary machine components
- user and maintenance handbook
- process and utility diagram
- electrical schematic papers
- qualification-related documents

Phase-II

This is the second phase, i.e. verification phase of qualification dealing with IQ, OQ, and PQ.

Installation Qualification (IQ)—The equipment/instrument and system are as installed/modified and comply with the approved design as approved and suggested by the manufacturer, according to IQ.

IQ refers to the collection of all operations required to ensure that equipment is delivered as designed and specified, that it is properly placed in the chosen environment, and that the environment is appropriate for the equipment/instrument [6–8]. IQ is largely for new/pre-owned equipment, as well as existing equipment that have not been qualified. The following are the typical sequences of events linked to IQ:

- description of the equipment
- delivery of equipment
- details of utility/facility/environment
- assembly and installation
- network and data storage
- verification of installation

Operational Qualification (OQ)

OQ is the documented verification that the installed or changed facilities, systems, and equipment work as planned within the specified operating ranges. During OQ, it is essential to have the Standard Operating Procedures (SOP) with the users. Machine book, manual, and maintenance plan with desired specifications and intervals are among the other documentary records to be kept for OQ. After the responsible person signs the qualification procedure, the Operational Qualification is considered complete [7].

The system's equipment and related services must function properly, and their operation must be certified in compliance with OQ's qualification protocol. It is necessary to identify and record critical operating parameters. The important variables should indicate the working conditions, with operational limitations ranging from extremes to extremes, as well as the condition that could lead to worst-case OQ criterion. Operational qualification entails checking the operation of all system elements, parts, services, controls, gauges, and other components. It is required to have a documented operational qualification report to show that the activities are satisfactory. Prior to use and implementation, standard operating procedures should be completed and approved. Operator training is also an important aspect of OQ, since they should be trained on how to utilize systems and equipment before they are used, with a written training report. The entire system and equipment should be released for routine use when the operational qualification has been completed, provided that all calibration, cleaning, maintenance, training, and related tests and results have been deemed to be acceptable [7]. The following are the minimal approval criteria for each component:

- Before beginning OQ, equipment must be calibrated.
- It must operate as described in SOP and according to the manual.
- It must meet all set parameters and challenges, such as temperature, pressure, flow rate, and speed control.
- It must operate safely and reliably throughout all intended and worst-case operating ranges; it must meet all set parameters and challenges, such as temperature, pressure, flow rate, and speed control.
- It must operate safely and reliably throughout all intended and worst-case operating ranges.
- It must operate safely and reliably throughout all intended.

Performance Qualification (PQ)

Performance Qualification is described as a documented verification that asserts and confirms that the equipment and system can operate effectively and reproducibly in accordance with the approved procedure and methodology and the product's preset specifications. PQ is the final and most important stage in the qualification process.

This step provides documented proof that the gadget produces the desired outcomes in a repeatable manner under production settings. If required, a clause of worst-case scenarios must be added in the Performance Qualification process to assure its quality. Equipment performance should be checked according to the performance qualification methodology. It is vital to keep the performance qualification report in written form in order to demonstrate satisfactory performance throughout time [9].

Following are the key parameters of PQ:

- performance checks
- testing frequency

The minimum Acceptance Criteria for PQ includes—

- Performance Qualification should be based on risk assessment and a worst-case scenario methodology.
- Each piece of equipment and system should follow the integrated system's instructions and meet all criteria, ranges, and limits for all physical, chemical, and microbiological characteristics.
- Performance Qualification should be done three times for each piece of equipment, with the first run indicating by chance, the second run indicating by accident, and the third run indicating consistency [9].

Preventive Maintenance and Repairs

When an instrument is used frequently, it is necessary to take specific precautions to avoid mistakes and to provide a repair mechanism for any existing or future problems or errors. If the instrument fails to pass the PQ test, it will need to be serviced or repaired [10].

For the majority of instruments, routine preventive maintenance should be scheduled and advised, and the appropriate PQ tests should be redone after the maintenance/repair to ensure the qualification is maintained [10].

Routine Practices for Operation, Calibration, Maintenance, and Change Control

There should be a process in place for frequent instrument maintenance, calibration, and operation. It is important to keep track of what you're doing and report on it. To be cost-effective, a qualification should be completed according to the premise "as much as needed—as little as possible". Existing documents can be utilized to minimize costs [10].

Other Miscellaneous Qualification

Aside from the four primary qualifications, DQ, IQ, OQ, and PQ, further qualifications are carried out along the process to ensure that the equipment functions properly. These includes

- Verification Qualification (VQ) is the documented verification that the equipment and system, as connected together, are still in good working order and provide the expected results and meet the user's expectations.
- Safety Qualification (SQ) is the documented verification that the equipment and system, as installed or modified, meet the process, facility, and personnel safety criteria.
- Maintenance Qualification (MQ) is the documented verification that the proposed equipment and system maintenance programme is appropriate for the intended purpose.
- Re-certification (RQ)

  Requalifying simply means going through the qualification procedure again. The verified confirmation that the systems are still working satisfactorily when joined together. When specific outcomes of relocation, considerable alteration, or ageing in the instrument exist, requalification is required. Requalification should be scheduled as needed, and the frequency is determined by numerous factors such as the study of calibration, verification, and maintenance findings. Periodic requalification should occur after changes, and the extent of requalification should be justified by a risk assessment of the change. Requalification is commonly thought of as a step in the change management process [11].

Acceptance Criteria

Acceptance criteria are an important and crucial part of the qualification process protocol. It is a critical component in determining if a procedure is successful or not. Few standards should be observed while defining acceptance criteria: they should indicate an appropriate level of accuracy and reliability. They should be observable and have a traceable reference standard that can be measured with an instrument or test procedure. They should be based on user/vendor-defined or industry-accepted specifications. They must be explicit, clear, and measurable when testing is completed [11].

Regulatory Requirements

In order to get the greatest performance with the specifications, equipment qualification QA and QC play an important role in laboratories and businesses. Equipment qualification guarantees that a piece of equipment or instrument is capable of producing accurate qualitative data or measurements. The ICH Q7A and Pharmaceutical Inspection Co-operation Scheme PIC/S under GMP for the pharmaceutical industry are guidelines that demonstrate that facilities, systems, equipment, and utilities are adequately qualified and maintained to ensure data and product integrity. There are no precise rules for determining the regulatory status of qualification; instead, GMP, ICH, FDA, and WHO collaborate to determine the specifications and norms that best suit the equipment's intended application [11].

Conclusion

Qualification is a word that refers to the process of transferring equipment from a vendor to an owner's site, starting with planning, executing, and obtaining the findings of an inquiry or analysis in order to check the equipment's capacity to work. The equipment must meet the acceptance requirements provided in the supplier's design qualification specification and guidelines. Qualification shall be completed on regular intervals and is relevant for the equipment, instruments, facility, and area before its use. There are several stages to the certification process, including the specification phase, which includes user requirement specifications (URS), factory acceptance tests (FAT), site acceptance tests (SAT), and design qualification. Installation qualification, operational qualification, and performance qualification are all part of the second phase verification. All new/existing equipment, facility, system, and instrument shall be qualified by the user with the assistance of the manufacturer/supplier of the equipment, instrument, system, and facility, as well as a qualified qualification team and engineering person. The system of the equipment should be qualified every 1–7 years from the date of the last qualification. Validation, qualification, and calibration are all interrelated phrases that refer to a group of operations rather than a single event. If a facility's development, operation, and maintenance are to continue to meet all regulatory standards, these three must be done on a regular basis. Qualification is essential since it contributes to the system's quality and performance.

References

1. Jain K, Bharkatiya M (2018) Qualification of equipments: a systematic approach. Int J Pharm Edu Res 8(1):7–14

2. Sharanya N, Ramasubramaniyan P, Jeyashree P, Srinag T, Palanichamy S, Solairaj P (2013) An overview on qualification of equipment—an ideal approach for equipment validation. Indo American journal of. Pharm Res 3(10):9013–9018
3. Bansal SK, Layloff T, Bush ED, Hamilton M, Hankinson EA, Landy JS, Lowes S, Nasr MM, St. Jean PA, Shah VP (2004) Qualification of analytical instruments for use in the pharmaceutical industry: a scientific approach. AaPS Pharm Sci Tech 5(1):1–8
4. McDowall RD (2009) USP [1058] analytical instrument qualification and the laboratory impact. J Spectrosc 22(6):315–322
5. Gowrisankar D, Abbulu K, Bala SO, Sujana K (2010) Validation and calibration of analytical instruments. J Biomed Sci and Res 2(2):89–99
6. Venugopal S (2015) A systematic approach to equipment qualification. European j biomed pharm 2(3):830–854
7. Smith P (2007) Trends in analytical instrument qualification. Bio-Process Int:30–38
8. Validation and qualification in the regulated environment (2016). Binder Whitepaper
9. https://www.pharmaceutical-technology.com/sponsored/pharmaceutical-qualification-and-validation/
10. https://www.pharmaguideline.com/2010/12/qualification-of-systems-and-equipment.html. Accessed Jan 2022
11. https://www.slideshare.net/Hardik_Mistry/pharmaceutical-qualification-validation. Accessed Jan 2022

Chapter 6
Approach to the Qualification Process

Abstract Assuring that something satisfies a requirement, standard, or set of requirements is known as qualifying. There is no clear definition for qualifying equipment, instruments, utilities, and facilities in Title 21—Food and Drugs, Part 820—Quality System Regulations in the Code of Federal Regulations. The qualifying process includes the following steps: installation qualification, operational qualification, process performance qualification, product performance qualification, and supplied material qualification. Equipment qualification is mostly used to confirm the accuracy of data. The equipment certification practices currently in use in the pharmaceutical industry are based on regulatory requirements, voluntary standards, vendor norms, and business practices. The certification process takes into account both the activity's nature and time. The four main methods for qualifying goods, services, and utilities are the following: Process performance qualification, Acceptable analytical practises, Qualification life cycle approach, and Quality risk management approach. These techniques place a higher priority on risk assessment and detection than they do on high-quality product performance. This chapter discusses the key criteria for operating certification of a number of sophisticated analytical devices, including the Karl Fisher titrator, dissolution apparatus, UV-spectrophotometer, IR, HPLC, and GC. The New ECA AQCG Guideline on Analytical Procedure Lifecycle Management's objectives are described in this chapter.

Keywords Qualification · Equipment description · Operational specification · Acceptance criteria · Risk management

Introduction

In order for something to be qualified, it must be made sure that it is installed correctly, operates correctly, and yields the expected results. Although qualification is a component of validation, individual qualification procedures do not, by themselves, constitute process validation. It involves getting products from producers or distributors, evaluating and testing them, and then determining whether they qualify

S. S. Shukla et al., *Pharmaceutical Calibration, Validation and Qualification: A Comprehensive Approach*, https://doi.org/10.1007/978-981-19-9002-1_6

as products. The terms qualification, verification, and validation are all linked [1]. The following provides definitions for the terms:

- A validation is an action, procedure, or occurrence that backs up or corroborates something based on reliable sources. The Code of Federal Regulations, Title 21—Food and Drugs, Part 820—Quality System Regulations, defines validation as "confirmation through examination and supply of objective evidence that the specified requirements for a specific intended use can be consistently attained". The Code of Federal Regulations, part 820.3(z).
- Several examples include Design Validation, Sterilization Validation, Test Method Validation, Software Validation, and Process Validation. When analysing a dynamic that is prone to unintentional alteration, validations are frequently used. Additionally, they are frequently larger projects requiring a great deal of subordinate qualifications (and/or verifications).
- Verification is the action or process of confirming something's reality or truth. According to the Code of Federal Regulations Title 21—Food and Drugs, Part 820—Quality System Regulations, it is certification by examination and presentation of objective proof that particular standards have been reached. The Code of Federal Regulations, part 820.3(aa).
- Two examples are the verification of designs and processes. Verifications are frequently used for assessing completed work that is static and not vulnerable to unintentional change.
- Assuring that something satisfies a criterion, standard, or set of standards is known as qualifying.

According to the Code of Federal Regulations, Title 21—Food and Drugs, Part 820—Quality System Regulations, there is no defined definition for qualification. Installation Qualification, Operational Qualification, Process Performance Qualification, Product Performance Qualification, and Supplied Material Qualification are all instances of qualifications. Qualifications are often a component of a larger validation project and have a lesser scope than validations. They are also less dynamic [1].

Approaches to Qualification Process

Equipment qualification is mostly used to confirm the accuracy of data. The equipment certification practises currently in use in the pharmaceutical industry are based on regulatory requirements, voluntary standards, vendor norms, and business practises. Because of this, pharmaceutical companies' approaches to laboratory equipment accreditation and their interpretations of the usually unclear criteria vary widely. Process validation, which ensures that the ongoing (dynamic) manufacturing process produces product that meets product/print specifications, includes Installation Qualifications, Operational Qualifications, Process Performance Qualifications, a Product Performance Qualification, and possibly Process Verifications [1]. Process

Validation's subset of Installation Qualification verifies that equipment was installed correctly (or possibly a Test Method Validation) [2]. The nature and timing of the exercise are two critical aspects that influence the entire process. The purpose of a qualification strategy is to make sure that the standards are met and to confirm them through inspection and the gathering of unbiased proof [2]. The qualification plan's steps are as follows:

Purpose

- scope
- equipment description
- operational specification
- acceptance criteria
- testing results
- discrepancies/corrective actions
- final report/conclusion

Main looms dealing with qualification of equipment/services/utility. They are—

1. quality risk management approach
2. acceptable analytical practices
3. qualification life cycle approach
4. process performance qualification approach [2]

Quality Risk Management Approach [QRM]

The risk can be estimated and evaluated using this method. Prior to implementation, the plan will identify areas that need additional analysis, adjustments, or redesign. Additionally, it will have measures to cut down on wasteful effort and duplication, lower risk, and shorten downtime for operators. Table 6.1 compares the methods used for facilities, utilities, and equipment.

Steps of QRM Approach

The steps in the quality risk management plan for the different requirements for facilities, equipment, and utilities are as follows:

- You can determine the impact by responding to the crucial determination questions. With a clear understanding of what "success" entails, assessments are conducted in accordance with a predetermined approach and/or protocol(s). The evaluation should be thorough but succinct.

Table 6.1 Comparison of quality risk management process for GMP facilities, utilities, and equipment

GMP facilities	GMP utilities	GMP equipment
User requirement and design review	User requirement, functional requirement, and design review	User requirement, functional requirement, and design review
Risk analysis	Risk analysis	Risk analysis
Validation strategy based on complexity	Validation strategy based on complexity	Validation strategy based on complexity
Commissioning	FAT (functionalities)	FAT (all configurations, facilities, capacity, functionalities)
Risk analysis review (initial and final)	Commissioning	Commissioning
Facility qualification	Risk analysis review (initial and final)	Risk analysis review (initial and final)
Traceability matrix	Qualification	Qualification
Final report	Critical parameter verification under operation conditions	Traceability matrix
–	Traceability matrix	Final report
–	Final report	–

- Write a user requirement specification (URS) document outlining all of the requirements of the process owner.
- Conduct a risk analysis that takes the needs of the user into account. In the final report, all deviations must be explained, and the conclusion must be concise and unambiguous (i.e. Pass or Fail).
- Choose the models that are used the most frequently, such as FMEA, Fault Tree Analysis (FTA), Hazard Operability Analysis (HAZOP), and Risk Priority.
- (Only applicable to GMP Critical [F/U/E]) Validation Strategy
- Choose a risk-based approach method. Who, when, and how will the results of the risk assessment be examined and confirmed?
- Develop a risk assessment test that includes functional requirement specifications (FRS) for equipment and utilities that contain control systems or software.
- Factory acceptance test (Equipment and Utilities only):
  - Drawings and specs, material of construction.
  - Alarm and interlocks, power failures, loss communication.
  - Environmental Health and Safety (EHS) Evaluation.
  - Critical instruments and major components.
  - Performance (cycles, setups, range, etc.).
  - Update the risk assessment and/or validation processes in light of the FAT results. A justification for the change in strategy must be given.
  - Carry out commissioning to make sure the installation was done correctly, to check that software updates (restricted to utilities and equipment) were sent, to validate that all documentation was received, and to make sure the utilities and equipment components work as they should.

   - Update the risk assessment and/or validation plan in light of the commissioning results. A justification for the change in strategy must be given.
   - The results of the risk assessment will be examined and verified. To demonstrate that the F/E/U is suitable for the use for which it is designed, carry out the following qualification, Bracing techniques or worst-case situations (Equipment and Utilities only) verification of crucial variables in practical circumstances:

     Monitoring of the environment.
     T and R, area classification, air flow pattern, and air changes.

- A review of the final risk assessment to ascertain how the identified risks were reduced, managed, or tracked.
- A final report to complete the qualifying process and a traceability matrix to keep track of all relevant requirements are also recommended.

Additionally, regular risk assessments should be performed. All FAT and commissioning-related tests and requirements must be justified, and at least one of the executable deliverables must test each requirement [3].

Recommendations

It is advised that the quality group get involved with the project right away. Detail-oriented requirements should be provided by the process owner and process engineer. Risk assessments and all deliverables must be amended and/or approved by quality [3] (Table 6.2).

Acceptable Analytical Practices

Acceptable Analytical Practices (AAPs) were developed by the Analytical Research and Development Steering Committee of the Pharmaceutical Research and Manufacturers of America to share information about how the pharmaceutical industry has implemented the CMC and Quality Guidance of the International Conference on Harmonization and international regulatory authorities. AAPs were developed to give people a way to learn from experts in pharmaceutical research and development and to raise knowledge of analytical practices that are in line with sound science and legal requirements. Sometimes unclear and subject to interpretation are the rules governing the qualification of laboratory instruments and equipment [3]. According to the requirements of good manufacturing practice, calibration is defined as "the calibration of instruments, apparatus, gauges, and recording devices at suitable intervals in accordance with an established written programme containing specific directions, schedules, limits for accuracy and precision, and provisions for remedial

Table 6.2 Description of parameters for qualification

Particulars	Process owner	Project engineer	Commissioning	QA validation	Vendors/ contractor
Project proposal/request	D	D		R	
User requirement specification [URS]	D	R/A		R/A	
Impact assessment	R/A	R/A		D	
Risk assessment [URS and process]	D	R/A		R/A	
Validation strategy	R/A	R/A		D	
Functional requirement specification	R/A	R/A		R/A	D
Design specification	R/A	R/A		R/A	R/A
Risk assessment	D	R/A		R/A	D
Traceability matrix development	R/A	R/A	D	R/A	D
Design review	R/A	D		R/A	R
Factory acceptance test development (FAT)	R/A	R/A	D	R/A	D
Factory acceptance test execution	R/A	R/A	E	R/A	E
Commissioning development	R/A	R/A	D	R/A	D
Commissioning execution/ verification	R/A	R/A	E	R/A	E
Risk assessment/review/ update	R/A	R/A	E	R/A	E
IQ	R/A	R/A	R	R/A	D/E
OQ	R/A	R/A	R	R/A	D/E
PQ	R/A	R/A	R	R/A	D/E
Individual summary report	R/A	R/A	R	R/A	D
Final risk assessment	R/A	R/A	E	R/A	E
Traceability matrix update	R/A	R/A	E	R/A	
Final report	R/A	R/A		D	

D develop, *E* execute, *R* review, *A* approval

action in the event accuracy and/or precision limits are not met". The good laboratory practice regulations state that "Equipment used for the generation, measurement, or assessment of data shall be appropriately tested, calibrated, and/or standardized".

Table 6.3 summarizes the crucial criteria for certifying the operation of various sophisticated analytical instruments and equipment used in laboratories [3].

Table 6.3 Key parameters for operational qualification of some equipment

Spectroscopic equipment	Chromatographic equipment	Other instruments
UV–visible spectrophotometer Wavelength accuracy Photometric accuracy Wavelength resolution	*HPLC* Peak retention time precision Peak area precision Accuracy of flow rate Accuracy of column oven temperature Detector linearity UV–visible detector linearity Gradient accuracy	*Dissolution apparatus* Temperature accuracy Accuracy of shaft rotation Equipment performance—USP calibrators Environmental vibration Control of paddle centring Control of distance of shaft to side of vessel
IR–NIR spectrophotometer Wavelength accuracy Wavelength resolution Photometric accuracy (for quantitative use)	*Gas chromatography* Peak retention time precision Peak area precision Temperature accuracy of column oven	*Capillary electrophoresis* Voltage stability Peak area precision
		pH meter Standardization
		Polarimeter Accuracy calibration
		Analytical balance Calibration
		Karl Fischer titrators Accuracy calibration

General Requirements to Set AAPs

- All equipment used to gather data for regulatory submissions must be qualified, maintained, and failures must be recorded as part of the programme. It is required to develop and document the equipment qualification programme.
- The use of a standard operating procedure (SOP) that details the entire programme.
- It is necessary to brand or tag equipment and maintain records. Equipment that is unqualified or performs differently than it should is marked.
- Every important operation, including failures, maintenance, certification testing, location, custodian, and others, is typically recorded in a logbook of some kind [3].

Equipment Maintenance

Some OQ testing should be done after equipment maintenance. The operational tasks affected by the question maintenance approach should be the only ones that are tested. Repetition of the OQ tests that assess the repaired or replacement component is typically required.

The best way to handle components that are prone to wear and require regular replacement is with a preventative maintenance programme. The qualification life cycle should include the establishment, documentation, and inclusion of preventive maintenance intervals. Intervals can be defined or altered based on the actual equipment qualification or maintenance history. Preventive maintenance is primarily carried out for business purposes because it is the most economical way to maintain equipment that requires frequent servicing [3].

Computers and Software Used to Control Equipment

According to current regulations, computers and software used to manage laboratory apparatus or process data must be validated. Systems that gather, process, or store data are subject to additional limitations described in 21 CFR 11. Although qualification is a part of validation, change management processes and additional documents are needed for validation. When numerous laboratory PCs are kept with the same setup of hardware and software, complete validation on a single representative unit is feasible. In any other identical combinations, only the IQ component of the certification would be necessary. Every time a configuration change is made because all of these machines must be maintained in the same change control system, revalidation is required [3].

Qualification Life Cycle Approach

There is no requirement for formal documentation of any equipment pre-purchase actions. The processes that result in the selection of equipment are referred to as design qualification, or DQ. When the equipment is delivered, the installation phase starts. The qualification cycle's installation qualification is the first step that needs formal documentation (IQ). Along with specialized installation tasks, labelling and making the equipment record to add the item to the inventory maintained as part of the programme are essential components of IQ. In rare circumstances, vendors may insist that customers install the system themselves, leading to shared documentation of IQ operations [3]. There is no requirement for formal documentation of any equipment pre-purchase actions. The processes that result in the selection of equipment are referred to as design qualification, or DQ. When the equipment is delivered, the installation phase starts. The qualification cycle's installation qualification is the first step that needs formal documentation (IQ). Along with specialized installation tasks, labelling and making the equipment record to add the item to the inventory maintained as part of the programme are essential components of IQ. In some circumstances, vendors may insist that customers install the system themselves, leading to shared documentation of IQ operations. Performance qualification is the testing of equipment using a particular technique or assay to ensure that the process is producing accurate data. PQ examples include control sample analysis and trending, system suitability testing, and method validation testing. The PQ increases the OQ by examining the precise methodology utilized [3].

Regulatory Environment

Today, validation is seen as a sequence of activities that take place throughout a person's lifespan rather than as a single activity. The qualification standards for utilities and analytical equipment/instruments have been modified by the ICH, GMP, and FDA. New sensor technologies are also utilized as part of PAT (Process Analytical Technology). Analytical methods require a lifecycle management approach to ensure that they are initially "fit for intended purpose" and continue to operate within their design capabilities throughout their working lives [3].

New ECA AQCG Guideline on Analytical Procedure Lifecycle Management (APLM)

The purpose of this ECA AQCG guideline is to provide:

- An outline and methodology for a lifetime approach to the qualification of analytical processes as part of a lifecycle approach that complies with the regulatory requirements and principles of the FDA process validation guideline.
- Descriptions and suggestions for the activities that make up the lifecycle's tiers.
- Contributions to the new Q14, the next ICH Q2(R1) modification, and assistance with the continuing creation of USP General Chapter 1220 on Analytical Procedure Lifecycle.
- Advice on how to use the lifecycle approach to enhance current practices. The industry's use of APLM in practise [3].

The key areas covered in the guideline include:

- principles of Analytical Procedure Lifecycle Management (APLM)
- prerequisites for the APLM

- guidance recommendations for the three stages of the APLM's analytical procedure
  - Stage 1: procedure design, development, and process understanding
  - Stage 2: procedure performance qualification
  - Stage 3: procedure performance verification
- six technical appendices

The Procedure Performance Qualification technique is then used, which helps the manufacturing and production departments of the pharmaceutical business perform better in terms of product quality and robustness [4].

Process Performance Qualification Approach

This is the core strategic, scientific, and statistical methodology that ensures the achievement of equipment qualification. It guarantees the longevity of the product. The analysis's objective is to gauge the degree of assurance in the product quality attribute measurements and contrast them with the standards. The number of process performance qualification (PPQ) batches needed to reach an informed and risk-based decision about the robustness of a product is best estimated using this method. Utilizing this strategy, batch variability is evaluated both within and between batches. The difference is that this technique separates out expected sources of volatility. Inter-batch sources account for some of the variation, whereas intra-batch sources are responsible for others [4].

The method described a statistical tool that calculates the number of batches, which should provide adequate scientific proof that the process is dependable and will regularly produce high-quality products. The method combines previously gathered product-specific data (such as data from Stage 1 batches produced for clinical trials, submission/registration, stability, and process scale-up/demonstration) with historical batch-to-batch process data (such as typical variability observed for this product/process type based on active content) to provide a predictive model [4].

The main goal is to identify the fewest batches for which a forecasted confidence interval of a product's critical quality attributes completely and readily falls within the necessary tolerances. That is, the number of batches that, in light of the information at hand, ought to provide adequate data to allow, following assessment, a statistically confident conclusion regarding the key qualities of the product. According to the principle of normal distribution, for a product to be in compliance with current standards, the tested mean of a product quality characteristic must be as close to the centre of the specification as possible and the standard deviation must be as little as possible [4].

The total or overall variability of a process can be expressed as the sum of the individual component variability. The overall variation is derived from sources of batch-to-batch, intra-batch, sampling, and analytical variability. This kind of variable is typical in processes [4].

The basic steps in attaining PPQ are as follows:

The Right Place to Begin

The manufacturer must prove that its manufacturing process is highly reliable and regularly yields the intended product (both API and drug product) within all standards through the gathering of pertinent data in order to obtain regulatory approval for the commercial sale of a new drug (purity, quality, potency, etc.). To successfully complete process validation, manufacturers must have a thorough grasp of their processes and the impact of process variances on product quality. Therefore, incorporating quality into the product and process from the start is the most efficient method to achieve process validation [4].

Defining the Risk

When Quality by Design (QbD) studies are completed, all of the action items listed in the initial Quality Risk Assessment (QRA) that was used to build the QbD studies are compared to determine how effective they were. These gaps in process knowledge are found and filled using this reconciliation, along with a more formal Failure Mode and Effects Analysis (FMEA), which is also used to determine the level of risk caused by any lingering potential process variations. The potential for significant effects is examined while changing several process parameters (charging rate, reaction temperature, reaction time, etc.). If a high risk is identified, further investigation is conducted to determine the range of variance that may be allowed, and mitigation plans are developed to prevent unacceptably high levels of risk [4].

The level of analytical readiness is also evaluated during QbD activities. When prioritizing tasks, it is important to take into account the urgency of the task in order to meet aggressive timelines for accelerated programmes (methods for the first step of a multi-step process), the depth of the required analysis (starting materials require fewer analyses than APIs), and the level of knowledge and information that is currently available (sponsor may have completed some to no method development work). Data gathering and other tasks necessary to address low-risk issues, like report preparation, can be carried out simultaneously with PPQ [4].

By this time, the sponsor must also have a clear regulatory strategy in place that specifies each nation where requests for clearance would be made. For example, the filing requirements for the validation of analytical procedures vary among nations. To meet the tightest requirements, the most effective strategy is to do all necessary work in advance [4].

Determining Regulatory Starting Material(s) and Campaign Strategies

Regulatory starting materials (RSMs) and campaign strategies must be established before the start of PPQ. The FDA must approve the RSMs used when cGMP manufacturing first began. Rarely, the sponsor firm or CDMO may not be informed prior to the final approval stage. As a result, there is a risk the agency could request a different RSM strategy, which will force the start of cGMP manufacturing earlier than expected. Risk mitigation can be performed by confirming additional process steps if there is any lingering uncertainty regarding the RSMs' consent. However, this choice must be made after carefully weighing the degree of risk against the time and expense required to validate the additional step(s) [4].

The needs of the sponsor frequently dictate the advertising strategy. The key choice is whether to validate and promote each stage independently or the entire process from start to finish. To minimize risk associated with a process or to create material to meet clinical demands, it may be beneficial to run through all of the steps once to confirm that the product meets standards, and then campaign the remaining batches for each step. However, it is challenging to explain a successful validation/ PPQ demonstration with fewer than three full manufacturing runs [4].

Pre-PPQ

Strong collaboration with the sponsor is necessary throughout all of the aforementioned steps in order to become ready for the process performance qualification phase. Process History Tables (PHTs) are updated along with Critical Process Parameters (CPPs), Critical Quality Attributes (CQAs), and Critical In Process Controls (CIPCs). According to DoEs and other laboratory and pilot plant investigations, the CPPs are the variables that have the biggest effects on the process and product quality. Control over these factors must be shown during the PPQ. A PPQ checklist is created and verified prior to the manufacturing of the PPQ batches to ensure that the pertinent documentation is in place and that any concerns or issues are discussed and resolved [4].

Establishing the Protocol

PPQ must be performed under cGMP conditions and in accordance with a set of rules. The written protocol must include information on the manufacturing process (including materials, CPPs, and other relevant factors), process controls, data to be analysed, sampling and analytical techniques, acceptance criteria for process performance, handling of deviations, equipment and facility/equipment design and qualification, operator training, and analytical method validation.

The uniformity of the final product must also be proven, in addition to process control. The regulations are murky on this point, but since they only apply to the drug product, where all doses must be the same, the API is also subject to the same requirements. Techniques for sampling and testing become extremely important in this situation because the API is not packaged in doses. Since all samples must fall within a statistically significant range, regardless of where they are collected, it must be demonstrated that a small sample truly represents the full batch.

The group must choose the measures for evaluation and the standards for homogeneity determination. Although meeting requirements could be sufficient, most criteria and data are determined and verified statistically.

The method for cleaning validation is defined by the type of equipment being used (shared vs. dedicated), as well as whether the PPQ process will be campaigned at each step or run from beginning to end. Cleaning validation occurs throughout the PPQ at each step or run from beginning to end for each batch [4].

Pre-approval Inspection

A strategic choice must be taken on when to submit the new drug application (NDA): after the protocol has only been designed or after the PPQ (i.e. validation) has been completed. Both methods are permitted per the rules. The PPQ is completed after the NDA is submitted and while it is being evaluated; hence, the first method can speed up the approval process. There is a danger, though, and unforeseen problems with finishing the PPQ could postpone a commercial debut.

The FDA is required by the 1992 Prescription Drug User Fee Act (PDUFA) to finish NDA reviews of typical projects within 10 months. If there is confidence in the manufacturing process (e.g. a QbD approach was used during the initial development phase) and thorough pre-PPQ planning has been carried out, this strategy may be useful for accelerating the approval process. However, this strategy is not advised for quick ventures because the deadline for finishing NDA reviews for these projects is only 6 months [4].

The Final Report

An accurate and comprehensive PPQ report that includes tabulated data, analytical findings, and an evaluation of adherence to the PPQ protocol must be submitted. Any unanticipated results and process variations must be recorded and justified.

It is also necessary to explicitly indicate any modifications to the process and/or controls that are necessary as a result of these findings. It is critical to state whether the data supports a successful demonstration of process qualification in the statement [4].

Conclusion

This chapter is primarily concerned with the methods used to qualify equipment and related facilities. In the pharmaceutical business and laboratories, qualification is a key aspect in the selection, procurement, installation, operation, and performance of any equipment, from the manufacturer to the vendor. It is the complete process of obtaining items from manufacturers or distributors, inspecting and testing them, and finally identifying them as qualified products. Nature (who, what, (potentially where), why, and how is the action being objectively proven) and time (when can the activity be begun and when must the activity be concluded) are critical for qualifying execution and implementation, as covered in this chapter. The regulatory agencies FDA, GMP, and ICH will emphasize the mandatory use of techniques for certification in the future.

References

1. Pazhayattil A, Alsmeyer D, Chen S, Hye M, Ingram M, Sanghvi P (2016) Stage 2 process performance qualification (PPQ): a scientific approach to determine the number of PPQ batches. AAPS PharmSciTech 17(4):829–833
2. FDA (2011) Guidance for industry, process validation: general principles and practices. US Food and Drug Administration
3. Sigvardson KW, Manalo JA, Roller RW, Saless F, Wasserman D (2001) Laboratory equipment qualification. Pharm Technol 25(10):101–108
4. A quality risk management approach for qualification and commissioning|IVT—GMP—Risk (2014)

Part III
Verification

Chapter 7
Objectives, Methods, and Role of Verification

Abstract Verification is the process of determining something's accuracy, truth, or validity. The verification process proves that the system or system component operates as anticipated and satisfies all performance specifications listed in the system performance specification, as well as functional and allotted baselines. Analytical methods can be validated and/or verified in order to show that they are appropriate for use and to guarantee the reliability, validity, and accuracy of the results. This chapter discusses the main differences and connections between validation and verification. The goals and purposes of the verification are also covered. On the basis of time and objective, the various verification techniques are examined. It is possible to employ a variety of verification techniques to make sure that a system or an element adheres to the design references or requirements. The four fundamental techniques for verification are inspection, demonstration, test, and analysis. Additionally, the planning, execution, and reporting parts of the verification process are covered, highlighting how crucial they are to any system. Peer reviews, walkthroughs, inspections, and other methods of practising software verification can help us avoid potential errors that would otherwise lead to software failure. Dynamic and static testing are the two types of verification testing. The V&V strategy must be applied across the whole lifecycle of the software development process. Finding flaws in a system and establishing whether or not the system is beneficial and employable in a real-world setting are its two main objectives. Software verification techniques that deal with error identification and remediation include debugging.

Keywords Verification · Validation · Confirmation · Static and dynamic · Demonstration · Inspection

S. S. Shukla et al., *Pharmaceutical Calibration, Validation and Qualification: A Comprehensive Approach*, https://doi.org/10.1007/978-981-19-9002-1_7

Introduction

Definition of Verification

Verification is the process of supplying unbiased evidence to support the assertion that predetermined standards have been satisfied. Verification is used to provide information and proof that the transformation was carried out in accordance with the chosen techniques, protocols, standards, or laws. The basis for verification is concrete proof, or data whose veracity can be demonstrated by factual findings attained by methods like measurement, testing, analysis, and computation, among others. Therefore, validating a system (product, service, enterprise, or system of systems) involves contrasting the realized characteristics or qualities of the product, service, enterprise, or system with its anticipated design features. Because it enables the programme to find faults in system components before they are integrated at the next level, eliminating expensive troubleshooting and rework, verification is a crucial risk-reduction activity in the implementation and integration of a system [1, 2].

Objectives of Verification (Fig. 7.1)

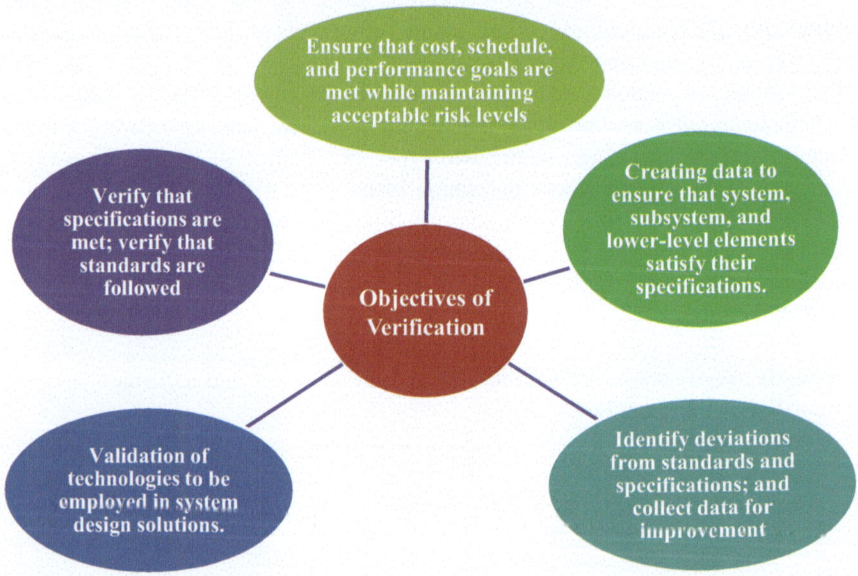

Fig. 7.1 Objectives of verification

Purpose of Verification

A method to validate each need must be defined and recorded during the operations of requirements analysis and functional allocation. Finding errors or defects that were introduced during the transformation of inputs into outputs is the objective of verification as a generic procedure. Any strategy can be used as long as it fits within the parameters of the project and the system's scope, includes some of the steps in the process outlined below, and is well-coordinated with other activities. The element that has been supplied uses generic inputs as a baseline. The design description of the system's internal and external interfaces, as well as the system requirements, is the inputs if the element is a system. The logical and physical architecture elements are the inputs if the element is a system. Generic outputs define the verification strategy, the specific verification actions, the verification techniques, the verification tools, the verified element or system, the verification reports, the issue/trouble reports, and the design change requests [1, 2].

The purpose of verification is (Fig. 7.2):

Difference Between Verification versus Validation

The references used to check an element's correctness and the acceptability of the effective correctness are the two most important differences between the verification and validation processes. It is crucial to keep in mind that, although verification is different from validation, it should be utilized in conjunction with it [2]. The difference is listed in Table 7.1.

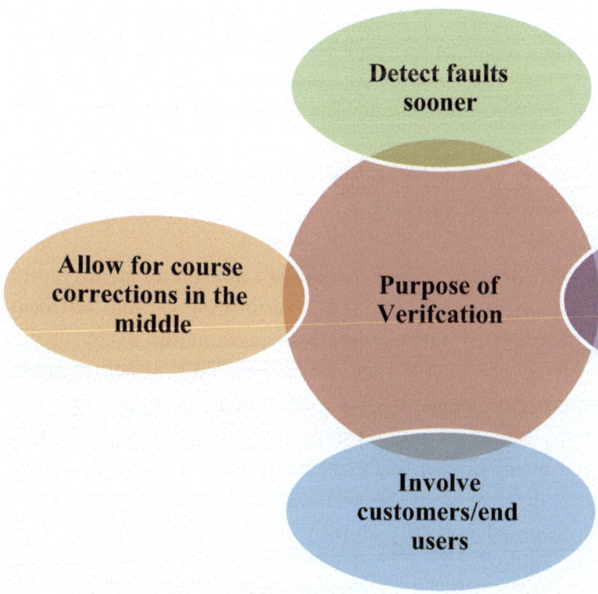

Fig. 7.2 Purpose of verification

Table 7.1 Difference between verification and validation

Verification	Validation
The software process conforms when it complies to the requirements	The procedure is what ensures that it satisfies the user's requirements
Verification is performed to make sure that the issue has been correctly fixed	To make sure one is working on the proper problem, one uses validation
The words verus, which means truth in Latin, and facere, which means to make or do, are the roots of the word verification	The word validation shares the same etymological origin as the word value and is derived from the Latin valere, which meaning to become strong
Verification is the process of demonstrating the accuracy or truth of a claim (a quality, a trait, etc.)	To validate something is to demonstrate that it contains the necessary components to result in the desired outcomes

Table 7.2 Examples of verified items

Items	Explanation for verification
Document	Verifying a document is examining how drafting conventions were used
Stakeholder requirement and system requirement	A stakeholder requirement or a system requirement is verified through the use of syntactic and grammatical rules as well as characteristics defined in the stakeholder requirements definition process and the system requirements definition process, such as necessity, implementation free, unambiguous, consistent, complete, singular, feasible, traceable, and verifiable
Design	Compare a system's logical and physical architecture components to the characteristics of the design process' results to validate the design
System	Verification is the process of comparing a system's actual qualities or features to its anticipated design attributes
Aggregate	To ensure the integration of an aggregate, each interface and interaction between implemented elements must be examined
Verification procedure	When evaluating a verification method, the use of a pre-established template and writing rules is examined

System Verification

System verification is a set of procedures used to validate any element, including system elements, systems, documents, services, tasks, requirements, and so forth. These kinds of operations are planned and executed throughout the system's life cycle. Every phase of the system's life cycle includes a cross-cutting action called verification. During the system development cycle, the verification process runs concurrently with the system definition and system realization processes. It is applicable to every action and every output that comes from the activity. Activities from each life cycle process and those from the verification process can cooperate. For instance, the integration process frequently employs the verification process [2].

The description is listed in Table 7.2.

Classification of Verification

Verification is classified on the basis of time and purpose. The types of verification/reviews are as follows:

- *In process review:* This is done throughout a particular point of the development cycle, like the design phase. In process review, finding errors in the work product and process is the aim. Through in-process reviews, faults can be found early in the process, when fixing them is less expensive.
- *Decision point and phase end:* As each stage of growth draws to a close, this stage involves a review of the processes and outcomes. Cost, schedule, risk, development progress, and readiness for the next phase are all taken into consideration while deciding whether to move forward with the development or not. This stage is also known as a milestone review.
- *Post-implementation review:* This is referred to as a post-mortem investigation. It is employed to enhance the procedure for developing software. It deals with the assessment and evaluation of products, including requirement compliance and predicted vs. actual development results. Post-implementation reviews are frequently carried out three to 6 months following installation [3].

Approaches to Verification

- Testing: Run the software to determine if it can generate errors as part of testing.
- Static verification: Examine the source code to find (specific) faults, i.e. take into account every possible path of execution.
- Inspection/review/walkthrough: A methodical collective study of the programme text to look for errors.
- Formal proof: Providing formal evidence that the programme text exactly complies with the programme specification [3] (Fig. 7.3).

Methods of Verification

To make sure that a system or an element follows to its design references or requirements, a variety of verification techniques can be applied. The four fundamental techniques for verification are inspection, demonstration, test, and analysis. Since each approach assesses a product's or system's requirements with escalating seriousness, the four techniques are hierarchical in character. Although there may be subtle differences on how the approaches are applied, these procedures are almost comparable to those used for validation. Validation is used to provide proof of conformity with (system and/or stakeholder) requirements, whereas verification is

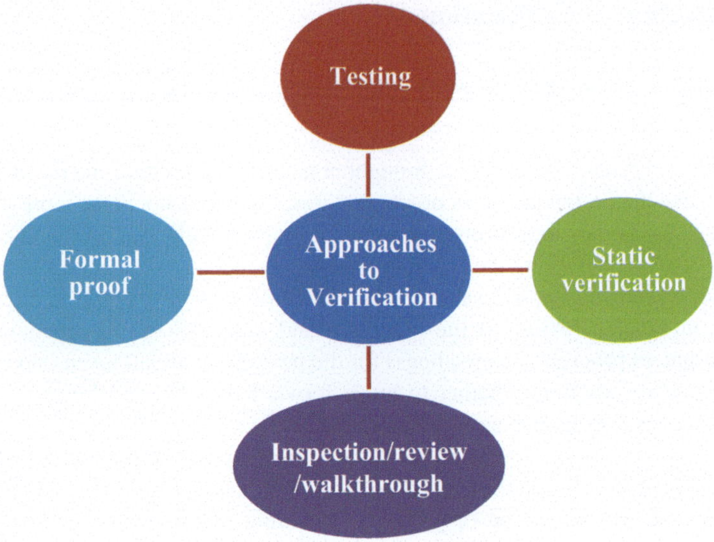

Fig. 7.3 Approaches to verification

used to look for flaws or problems. The techniques to some types of verification are described here [3].

Inspection/Examination

The verification relies on visual or dimensional examination of an element; it makes use of simple handling and measuring techniques or relies on human senses. Common means of inspection include the use of sight, hearing, smell, touch, and taste, as well as straightforward physical manipulation, mechanical and electrical gauging, and measuring. Tests or other stimuli are not necessary. Inspection of a system, part, or subsystem visually. It is typically used to verify certain physical design elements or pinpoint a specific manufacturer. The technique is used to confirm characteristics or attributes that are best determined by observation (e.g. paint colour, weight, documentation, listing of code). It is the process of doing a non-destructive inspection of a system or product using one or more of the five senses. Measurements and straightforward physical manipulation could be involved. Visual inspection of the equipment and study of drawings and other significant design data and processes should be used to guarantee conformance with features such as physical, material, component and product marking, and craftsmanship [3].

Demonstration

To guarantee that the outcomes are as intended or anticipated, it is necessary to manipulate a system or product in the way that it was intended to be used. This method was used to compare the submitted element's accurate operation against operational and observable features without using physical measurements (no or minimal instrumentation or test equipment). Field testing is another name for demonstration. It often consists of a number of tests that the provider has selected to show that the element responds appropriately to stimuli or that operators can carry out their assigned tasks while using the element. The outcomes of the observations are contrasted with the predicted or anticipated outcomes. Demonstration may be appropriate when needs or specifications are expressed in statistical terms (e.g. mean time to repair, average power consumption). Demonstrating that a requirement can be satisfied by a system involves using the system, subsystem, or component operation, to put it simply. It varies from testing in that it does not gather a lot of data and is often used for basic assurance of performance capabilities. Demonstration refers to the execution of operations at the system or system element level where visual observations serve as the primary method of verification. Demonstration is used instead of quantitative assurance for requirement verification [3].

Test

A product or system is tested using a controlled and predefined set of inputs, data, or stimuli to make sure it will produce a very specific and predefined result that is stated by the requirements. The operability, supportability, or performance capabilities of the submitted element are quantitatively confirmed when put under actual or simulated controlled conditions. Testing typically makes use of specialized test apparatus or equipment to provide exact quantitative data that can be evaluated. The use of a system, subsystem, or component activity to collect particular data in order to validate performance, check for errors, or give adequate data for further analysis. The system design must ultimately be verified by testing, a thorough quantified method of verification. A test is a process for learning about how a piece of equipment operates and how its functional aspects work in a controlled setting. The data is then used to evaluate quantitative attributes [3].

Analysis

It involves the use of simulations, calculations, and testing tools to validate a system or product. Based on verified test results from a sample set or by integrating the results of various tests to get a new conclusion about the product or system, analysis enables someone to make predictions about a product's or system's typical performance. Non-destructive tests are widely used to extrapolate the failure point in order to predict the breaking point or failure of a system or product. Technique for establishing theoretical conformity based on analytical evidence produced without any intervention on the submitted element by logical reasoning (including the theory of predicates), modelling, and/or simulation under specified conditions. This strategy is used when real-world testing is not feasible or cost-effective. Analysis is the process of using established analytical techniques (such computer models) to comprehend or clarify the behaviour or operation of a system component. Test data analysis or a review and analysis of design data should be used as necessary to validate requirements. A design's conformity to its requirements is estimated using calculated data or data generated from testing of lower level components or subsystems using mathematical modelling and analytical techniques. It is typically utilized when a physical prototype or product is unavailable or not economically viable. The analysis employs both simulation and modelling [3].

Verification Process

The three steps in the verification process include Planning, Execution, and Reporting (Fig. 7.4).

Step 1: Verification Planning

Verification planning is performed at each level of the system under development. The following activities describe the development of a verification plan:

- *Verification Method and Level Assignments*: The relationships between the verification level and the given requirements method are specified. For each level of

Fig. 7.4 Verification process

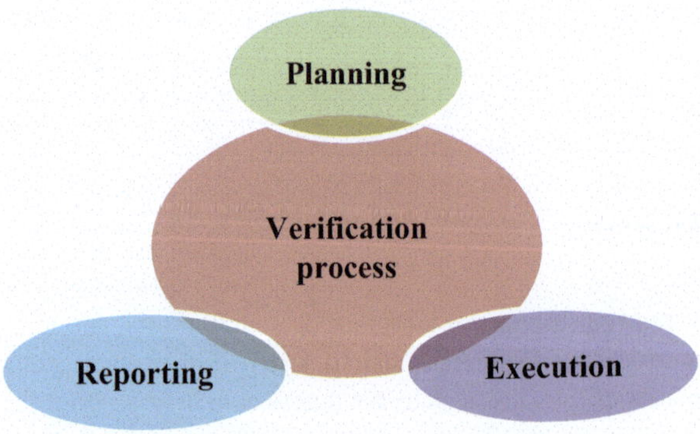

the architecture, this activity often produces a Verification Cross Reference Matrix, which is used to specify the verification activities. The level of verification is determined by the level of the demand (e.g. system level, subsystem level). Verification activities include, among others, analysis, inspection, demonstration, and testing. (See below for more details.) It is necessary to consider the verification technique as a potential risk. The application of inefficient techniques may lead to incorrect verification.

- *Verification Task Definition:* Lists all verification tasks, each of which responds to one or more conditions. The test engineer must have a solid understanding of how the system will be used and its related environments in order to create outstanding verification jobs. The requirements, functional, and physical designs that make up the integrated architecture are crucial components for the test engineer. A family of verification jobs is built to evaluate the functional, performance, and constraint requirements when combined with the physical architecture. The functional architecture is used to make it easier to develop functional and performance tests.
- *Verification Configuration Definition:* Outlines the technology configuration, resources, including people and settings, and need for a certain verification task. This might have components that mimic the system's external interfaces in order to execute tests, such as hardware or software.
- *Verification Scheduling:* Specifies the verification task schedule, whether the activities are carried out sequentially or concurrently, and the enabling resources needed to fulfil the tasks [3].

Step 2: Verification Execution

The accomplishment of a given verification task with aiding resources. Whether from a test, analysis, inspection, or simulation, the results of the verification task are documented for compliance or non-compliance, together with supporting information [3].

Step 3: Verification Reporting

Reports the combined findings of the carried-out verification plan and attests to the safety and environmental friendliness of the materials utilized in system solutions [3].

Software Verification

- The capacity of the programme to comply to its standards is referred to as "verification". Software verification is the process of looking at any software in order to find faults. Verification is the process of making sure software accomplishes its goals without mistakes. It is the process for figuring out whether or not the product being generated is accurate.
- A document can be reviewed using the first stage of software development, known as software requirement and analysis, which produces the SRS document.
- Despite the fact that verification is more likely to be successful than validation, it could still find some defects that were difficult to find during the validation stage. At the same time, it makes it possible for us to find software defects at the earliest possible time.
- Verification makes certain that each stage of the software development process results in the anticipated outcome [3].
- Validation ensures that newly created or changed software satisfies all functional and other specifications [3].

Methods of Software Verification

Peer reviews, walkthroughs, inspections, and other methods of practising software verification can help us to stop potential faults that would otherwise lead to software failure [3].

Peer Reviews/Informal Reviews

The easiest and most informal method of assessing papers, programmes, or software in order to find faults throughout the verification process is peer review. In this approach, people are given documents or software programmes to evaluate with the goal of getting their feedback on the quality of our output as well as finding any weaknesses in the programme or document. Using this method, the reviewers may also produce a brief report on their findings or observations [3].

The activities that are involved in this method may include:

- SRS document verification
- SDD verification
- program verification

Advantages

- Without spending a lot of money, positive outcomes could be anticipated.
- It is quite effective and substantial in character.
- More cost-effective.

Disadvantages

- If the reviewer is insufficiently knowledgeable, the result will depend on their abilities [3].

Walkthrough/Semiformal Reviews

Walkthrough is a more informal and structured form of verification technique than peer review. The creator of the software document walks a group of two to seven people through it in a walkthrough. There are no preparation requirements placed on the contestants. The presenter is in charge of setting up the meeting. Copies of the document are distributed to each participant (s). During the walkthrough meeting, the author introduces the material to familiarize everyone with it, and everyone is welcome to ask any questions they may have [4].

Advantages

- It might help us find any potential problems.
- It can also be used to distribute papers to others.
- Knowledge exchange.

Disadvantages

- The author could overemphasize certain aspects of his or her field of interest and omit other, more important details.
- There aren't many issues, however it is quite expensive.

Inspections/Formal Review

Inspections, which are commonly referred to as inspections, are the most organized and formal method of verification. A group of three to six individuals is assembled, and a neutral moderator takes charge. Everyone in the group participates honestly, willingly, and in accordance with the rules for carrying out such a review. Everyone may be given the chance to express their opinions, potential flaws, and important areas. The moderator's key recommendations are incorporated into a final report after the conference [4].

Advantages

- It can be quite helpful in identifying probable weaknesses or errors in papers like SRS, SDD, and other similar ones.
- Effective yet prone to errors.
- Critical inspections may also help in the discovery of problems and the modification of these documents, preventing the propagation of a flaw in the software development life cycle process [4].

Disadvantage

- They demand discipline and need time and effort.
- It is more expensive and requires the employment of qualified testers.
- It costs a lot of money and calls for the involvement of professionals.

Rules of Review

Flaws are discovered, not fixed, and everyone on the evaluating team is responsible for the review's conclusions. The product is reviewed, not the maker [4].

Categories of Verification

Dynamic Testing

- Involves choosing a group of test cases and test data, as well as running a system or component.
- Determining output test results using input test cases.

Functional, structural, and random testing are the subcategories of dynamic testing.

Functional Testing

- Defines and tests each of the system's functions according to the documentation outlining the fundamental requirements.
- It is a "Black Box Test", meaning that the tester is not expected to be familiar with the system's fundamental programming or implementation.
- It makes use of Test Cases Intended to Examine Particular System Features.

Structural Testing

- It is White Box Testing, which presupposes total familiarity with how the system is implemented.
- Tests that verify the operation of specific components are created using knowledge of the internal organization of the system.
- Like functional testing, it uses test cases created to investigate certain system features.

Random Testing

- Uses a collection of all possible test cases to choose test cases at random.
- Uses randomly determined test conditions to find faults not found by other systematic testing techniques.
- The process of constructing input test cases with all potential input values is known as exhaustive testing.

Static Testing

- Do not involve operation of the system or component.
- Involves both manual and automated testing techniques.

The sub-category of static testing is:

- Consistency techniques: analysis of consistency.
- Measurement techniques: measurement of some property of the program [5].

Static and Dynamic Verification

Examinations of the software concerned with identifying issues through examination of the static system representation (static verification). To supplement it, tool-based document and code analysis may be used. A product is put through its paces during software testing to evaluate how it responds (dynamic verification). Test data are used to run the system, and the outcomes are seen.

Verification and Validation Goals

Verification and validation should provide assurance that the programme is fit for its intended use.

- Software functionality: Compared to prototype systems, for example, safety-critical systems require a much higher level of assurance that the system is fit for purpose.
- User expectations: Users have low expectations for software and are prepared to put up with minor system problems (although this is decreasing).
- Marketing environment: The marketing environment should be taken into account, along with the timeline for launching the product. Products that cost less often have problems [5].

Types of Testing

- A sort of testing called defect testing is used to identify weaknesses in a system. A system's weaknesses are discovered by a successful defect test.
- The term "statistical testing" refers to tests that are intended to simulate the frequency of user inputs. It calculates a system's reliability [5].

Testing and Debugging

Debugging and testing for defects are two distinct processes. Verification and validation are performed in order to check for errors in programmes. Errors are found and fixed during the debugging phase. In order to identify a system error, debugging requires formulating a hypothesis about a program's behaviour and verifying it.

V&V Planning

- Careful planning is needed to make the most of testing and inspection procedures.
- Planning should start as soon as possible after development.
- The balance between testing and static verification should be specified in the plan.
- Test planning focuses more on establishing guidelines for the testing procedure than it does on outlining specific product tests.
- The system specification and design should serve as the foundation for the test plans [5].

Conclusion

Verification, in simple terms, is the process of confirming the accuracy of a hypothesis or fact. The act of reviewing, inspecting, or testing a product, service, or system in order to establish and record that it fulfils regulatory or technical requirements is known as verification and validation in engineering and quality management systems. Static verification is focused with analysing the static system representation in order to find issues, whereas software testing is concerned with exercising and observing product behaviour as an indicator of dynamic verification. Planning, Execution, and Reporting are the three processes in the verification process. Testing, static verification, inspection/review/walkthrough, and formal proof are examples of verification methods. Every firm should employ verification as a crucial procedure to assure system dependability and reproducibility. Validation and verification work in tandem and are inextricably linked. The quality of software can be improved by employing the V&V process to verify and validate it on a regular basis. Verification is critical to a software system's operation, accuracy, and output outcomes. Defect testing and statistical testing are two types of verification testing that are commonly used to estimate dependability. Verification and validation should provide assurance that the programme is fit for its intended use. Defect testing and debugging are two different things. The purpose of verification and validation is to determine whether or not a programme has any faults. Debugging is the process of detecting and correcting mistakes. The "Clean room" technique, which was developed with the idea of avoiding flaws rather than removing them, is the most recent contribution to software verification. More of these emerging technologies may help save time and money by improving efficiency and reproducibility, as well as providing long-term features for software verification.

References

1. MITRE (2011) Verification and validation. In Systems engineering guide. Accessed 11 Mar 2012
2. http://www.csc.liv.ac.uk/~coopes/comp201. Accessed 26 Jan 2022
3. Buede DM (2009) The engineering design of systems: models and methods, 2nd edn. Wiley, Hoboken
4. INCOSE (2012) INCOSE systems engineering handbook, version 3.2.2, INCOSE-TP-2003-002-03.2.2. San Diego, CA: International Council on Systems Engineering (INCOSE)
5. ISO/IEC/IEEE (2015) Systems and software engineering - system life cycle processes, ISO/IEC/IEEE 15288:2015. Geneva: International Organization for Standardization (ISO)/International Electrotechnical Commission (IEC), Institute of Electrical and Electronics Engineers

Chapter 8
Improvement in Software Quality

Abstract At all levels, software quality is now a crucial aspect of project management. If the software is of better quality, project delivery will be more efficient financially and operate more effectively. Increased productivity results in the creation of better software. Quality assurance, quality control, and testing are the three main components of software quality management, which is a broad term. One of the main methods in software testing best practices is PDCA, or plan, do, check, and act. It is an efficient control mechanism used to control, govern, monitor, regulate, and restrain a system. This chapter outlines the essential component, seven doable ways, and signals for enhancing software quality. Two models for enhancing software quality are the Process Reference Model and the Content Reference Model. Ten steps for improvement in software quality have been discussed in this chapter.

Keywords Software · Quality · Testing · Quality control · Quality assurance · Retrospection

Introduction

A software product is developed to satisfy specific customer requirements, but frequently it turns out to be a defective product for a variety of reasons, including incorrect requirements, communication or understanding gaps, scheduling problems, lack of sufficient technical knowledge, or system personnel with low skill levels. This makes the software products susceptible to bugs, flaws, or mistakes. Software testing is crucial for maintaining the calibre of software products and avoiding or preventing these kinds of problems [1].

Implementing efficient QA management that offers tools and processes for developing bug-free solutions is the best way to produce high-quality software. Quality assurance, quality control, and testing are the three main components of software quality management [1].

S. S. Shukla et al., *Pharmaceutical Calibration, Validation and Qualification: A Comprehensive Approach*, https://doi.org/10.1007/978-981-19-9002-1_8

Software Quality Assurance (SQA)

It is the area of quality management where a set of deliberate organizational actions are included. The goal of these measures is to enhance the software development process by adding quality standards to prevent mistakes and problems in the final output [1].

Software Quality Control (SQC)

It is the area of quality management that consists of a series of tasks aimed at meeting standards for quality. Prior to delivery, software products are certified for quality by product-focused activities known as quality control (QC). Software quality assurance regulates the software quality control process [1].

Testing

It is the fundamental process for identifying and resolving technical problems in software source code as well as evaluating the usability, performance, security, and compatibility of the final result. It is not only the primary component of quality assurance, but it is also a crucial step in the creation of software. Three things can be done, particularly in software development, to enhance our software product [1]. These are as follows:

- Preparation Quality Assurance
- Preventive Quality Assurance
- Reactive Quality Assurance

Areas for Step Up in Software Quality

Improving software quality requires software testing. Software testing is the procedure used to determine whether the programme complies with the necessary specifications. Continuous improvement should be the goal of the process. These methods are picked out and used. The most commonly used method is the Deming wheel (PDCA cycle). Maintenance expenditures are decreased by enhanced test process quality [1].

Fig. 8.1 Deming wheel

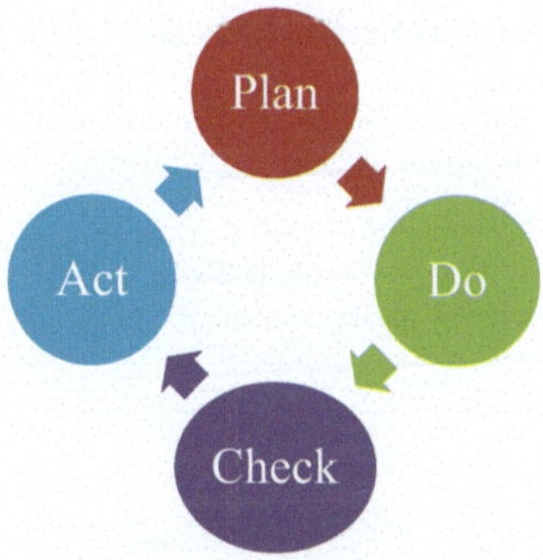

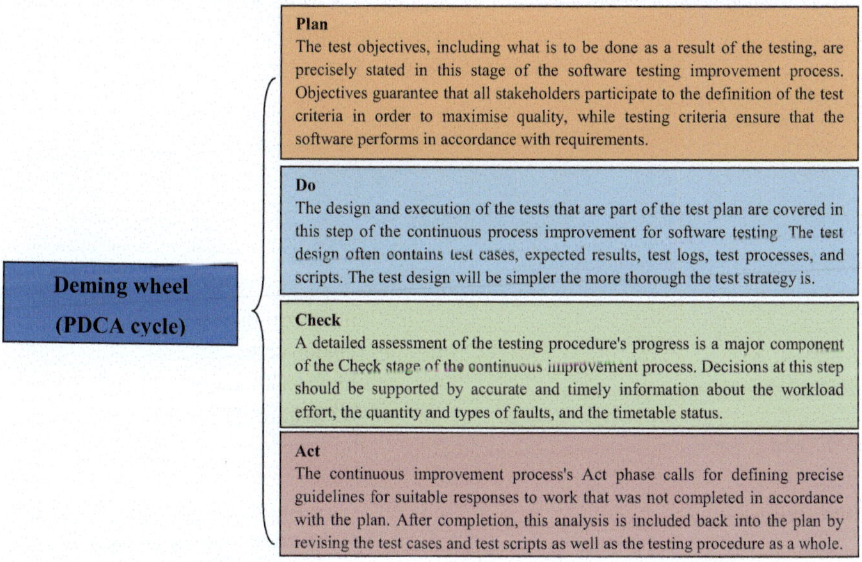

Fig. 8.2 Description of PDCA cycle

Deming Wheel (PDCA Cycle) (Fig. 8.1)

Implementing efficient and timely QA testing best practices that provide sturdy tools and techniques to produce flawless products is the greatest way to ensure high-quality software. Software life cycle testing basically means that testing is a continuous activity that happens in conjunction with the development cycle. The software testing process should be initiated early in the application lifetime and integrated into the creation of the programme itself [1].

Plan, Do, Check, and Act (PDCA) is a powerful control mechanism used to control, govern, monitor, regulate, and limit a system and is one of the best practices for software testing [1–3] (Fig. 8.2).

Key Rudiments and Meters for Progress in Software Quality

The four recommendations will increase testing effectiveness and software quality (Fig. 8.3).

The two indicators which make software quality tangible are depicted below (Fig. 8.4):

Fig. 8.3 Key rudiments and meters for progress in software quality

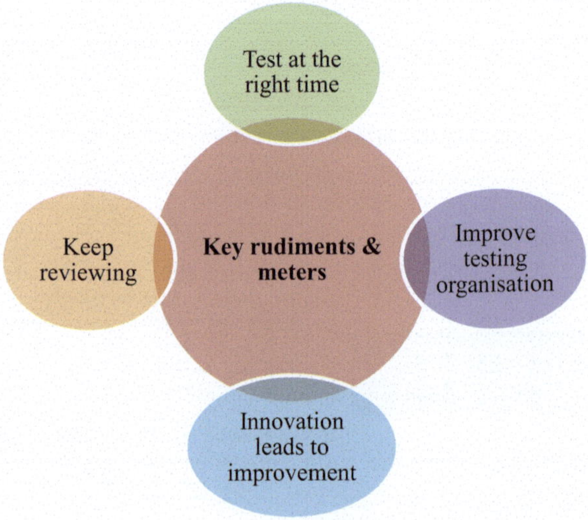

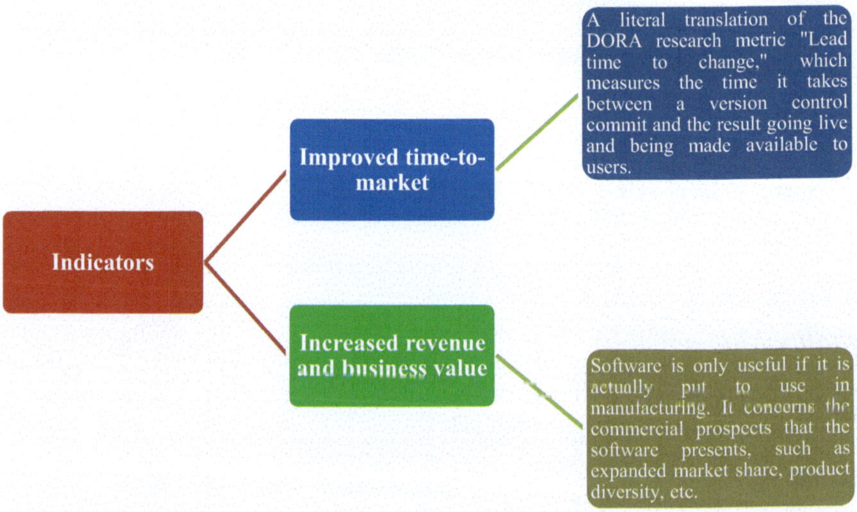

Fig. 8.4 Indicators which make software quality tangible

Fig. 8.5 Key elements to improve software quality

Software quality can be greatly increased by using five essential components. These components include a process, people, and technological viewpoint. In experience, the finest outcomes come from combining these viewpoints [1–3]. These are as depicted in Fig. 8.5.

The best possible ways for software testing with improved quality are as follows:

Devising a Plan and Defining Strategy

Effectual planning necessitates the conception of quality management and test plans for a project [1–3].

Quality Management Plan

Defines a clear and acceptable level of product quality and describes how the project will achieve the said level. The main components of a quality management plan are given in Fig. 8.6:

Fig. 8.6 Quality management plan

Test Strategy

An effective strategy should have a thorough introduction, a general plan, and testing specifications. A test strategy's primary elements include (Fig. 8.7):

Scenario Analysis

Regardless of how thorough a test plan is, issues will inevitably arise and carry over from one test phase to the next. Therefore, post-project and in-process escape analysis is essential for guiding test improvement [1–3].

Test Data Identification

- **Automated testing**
  The test early and test often strategy is often followed by continuous testing and process improvement. Automated testing is an excellent way to quickly assess the quality of an application.

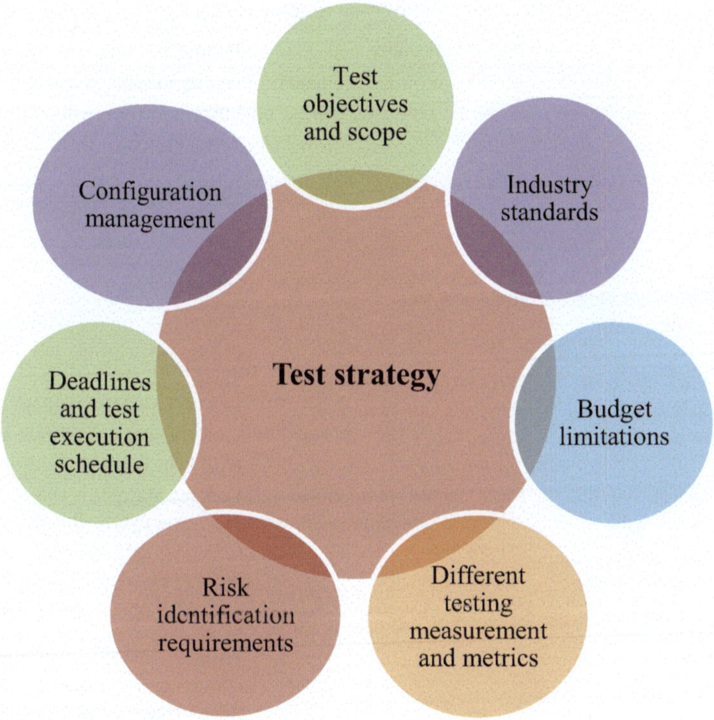

Fig. 8.7 Test strategy

- **Pick the right QA tools**

  It is crucial for testers to use the best testing tools based on the needs and goals of the test. The most often used tools are those from Jenkins, Selenium, GitHub, New Relic, and others.

  The most effective methods for QA improvement mostly involve designing the complete process for automated QA testing, selecting the appropriate tools, integrating QA with other tasks, setting up a stable testing environment, and carrying out continuous testing [1–3].
- **Robust communication between test teams**

  Ongoing communication always results in continuous improvement. Consideration of frequent communication between teams whose activities overlap during an active product development cycle is an excellent technique, especially in software testing best practices. This makes sure that they are exchanging ideas, observations, and solutions with one another [4].

The seven pragmatic steps for improvement in software quality are discussed in Table 8.1.

Table 8.1 Seven pragmatic steps for improvement in software quality

S. No	Steps	Impact on quality	Benefit
1.	Define quality to match the needs	Meet business requirements; achieve a satisfying user experience	Ability to achieve quality is improved because the application development team is not charged with unrealistically perfect expectations. Rather, it is chartered with a definition of quality that fits the given time, resource, and budget constraints
2.	Broadcast simple quality metrics	Reduce defects	Highly visible metrics keep quality top of mind for the entire team and expose when efforts fall short
3.	Fine-tune team/ individual goals to include quality	Meet business requirements; achieve a satisfying user experience; reduce defects	Team members perform according to their incentives, making quality improvement part of their goals reinforces desirable behaviour
4.	Get the requirements right	Meet business requirements; achieve a satisfying user experience	Less rework means less retesting and fewer cycles, this greatly reduces the overall effort
5.	Test smarter to test less	Reduce defects	A focus on testing the most crucial and at risk areas ensures that they receive the lion's share of test resources and that any bugs that slip through are likely to be confined to the least-important features
6.	Design applications to lessen bug risk	Reduce defects	Simpler, cleaner designs result in code that is simpler, cleaner, and easier to test and rework—which means that the code will have fewer bugs and that those bugs will be easier to diagnose and repair
7.	Optimize the use of testing tools	Reduce defects	Automation frees resources from mundane testing to focus on the highest priority tests and increases test cycles' repeatability

Models for Improvement in Software Quality

There are two models as listed below (Fig. 8.8):

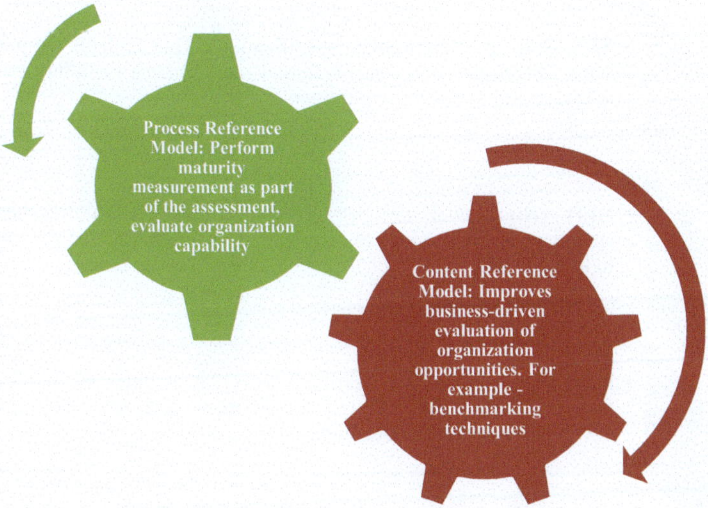

Fig. 8.8 Models for improvement in software quality

Steps for Improvement in Software Quality

Test Early and Test Often with Automation

Software testing is now an essential step in the life cycle of software development. Many businesses pick test automation as smart work in order to lessen the manual labour-intensive nature of testing. Automation capabilities eventually go beyond time reduction to speed up, cover all tests, and most significantly, optimize QA expenses. Therefore, test automation is favoured over manual testing in order to find a substitute with the highest achievable performance or cost-effectiveness to obtain the best possible result with the least amount of money spent [4].

Requirement Specification Document Availability

Some of the key requirement documents include (Fig. 8.9):

Testing Team Concern in Requirement Discussions

Clear and effective communication between all members of the design, development, and testing teams is one of the core elements of creating a successful project. In order for the testing team to refine the next job, they should be present at all

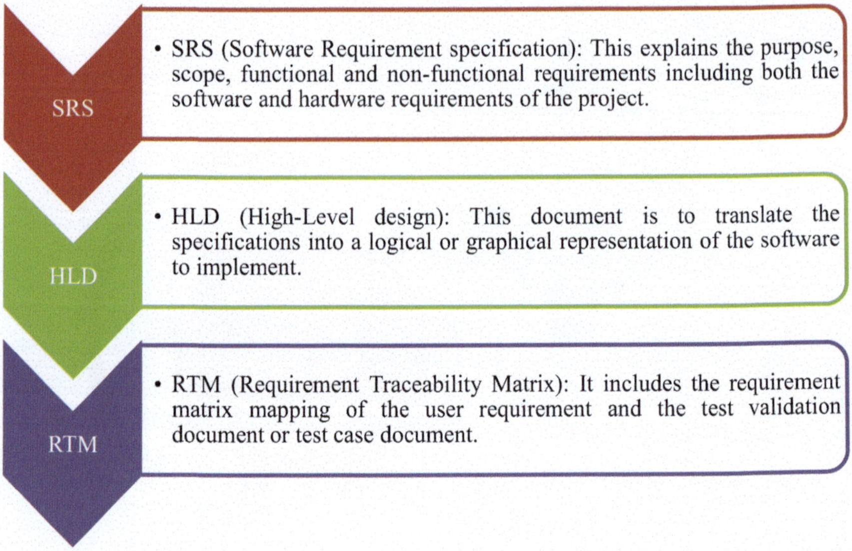

Fig. 8.9 Specification document availability

significant meetings and design meetings, including application designs and requirement defining sessions [5]. The concern therefore is as follows (Fig. 8.10):

Clear-Cut Scope

- Since the bulk of software development in the IT industry employs the agile methodology, the client rarely delivers a complete or well-defined scope and frequently alters the requirements throughout the development cycle.
- The development and testing teams argue as a result, and the results occasionally turn out differently than expected.
- Enhancing the software testing process: The testing team should have a clear understanding of the requirements before starting software testing, and the scope should always be known. In actuality, doing so will always lead to better results.
- The type, degree, and extent of testing required can be more easily determined if the project's full scope and aim are known [5].

Test Planning and Execution

The whole testing process, including developing requirements, methods, corporate standards, documentation, functionality descriptions, and potential dangers during

Fig. 8.10 Testing team concern

testing, are all covered in this phase. In order to produce a high-quality product, test planning is a full project that is divided into the following crucial tasks. To carry out the testing requirements within those procedures, a high-level description or document of the test procedure must be prepared. The testing team adheres to the strategy outlined in these publications. The test manager creates the test strategy document, which is static and does not change regularly. Test Lead must create a master and detailed test plan that is generated from the SRS document after creating a test strategy document. Test Case Design is the process of formalizing all requirements talks into documents like test cases, test scripts, and test scenarios. Or, to put it another way, test cases are a set of procedures that a tester uses to determine whether or not a software product satisfies all the requirements by contrasting the actual result with the anticipated outcome [5].

Test Case Review

Test case review plays an important role in the software development life cycle in any organization as the ultimate goal of the customer is to get a product "which is defect free" and should meet all the specified requirements. The main purpose of reviewing test cases: to estimate completeness, increase test coverage and correctness of the analysed requirements, and most importantly "No gap between requirement understandings" thereby improving the product quality [6, 7].

Regression Test Planning

The development team typically releases changed builds to the testing team to validate faults after making the necessary modifications in the software coding to remedy the issues. Sometimes, even a minor modification in the coding might have a significant impact on the software's other, unaffected portions.

Regression testing should always be planned in order to assure the management team, developers, testers, and clients that the new feature is not having any negative effects on any of the existing functionality and to make sure that any new problems are not being revealed in those functionalities that have not changed [6, 7].

Importance of Regression Testing (Fig. 8.11)

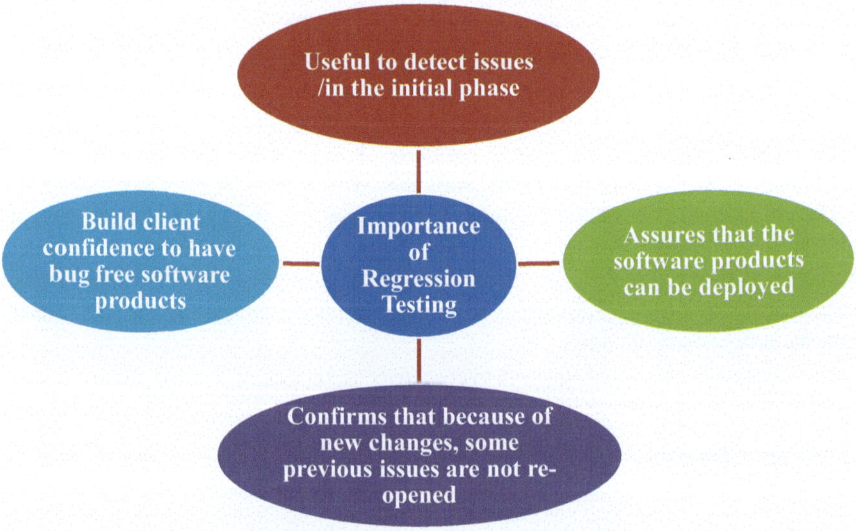

Fig. 8.11 Importance of regression testing

Different Ways to Perform Regression Testing

Every time new functionality is added, an existing product has a bug that has to be fixed, current functionality is changed, or existing features are removed, regression testing is necessary [8]. These code alterations may cause the system to develop a new flaw and behave improperly. The various methods for conducting regression testing include:

- retesting of complete test suit
- selection of regression test cases
- prioritization of test cases

Test Data Management and Reporting

The practice of managing testing activities, such as allocating testing resources, estimating, planning, and strategizing test efforts, as well as monitoring test progress, reporting, and control, is known as test management.

Test management is a powerful tool for streamlining the software testing process and delivering high-quality software products [8].

Retrospection After Every Sprint

A software development team will hold a retrospective meeting at the conclusion of a sprint to review and evaluate success and failure, as well as to make new plans for future improvements for subsequent sprints. Retrospectives are conducted after each sprint to provide teams an opportunity to continuously improve their performance and all other activities that are involved, not only the software testing process.

Start with quality controls in place and emphasize the value of quality control throughout the whole software development process.

The process of quality control begins from the outset and continues all the way to delivery. QA should be present at all times during the software development cycle. The project team's provision of quality assurance is the governance that fosters trust in the overall software quality. Assurance testing supervises and confirms that the procedures used to deliver results have been monitored and are operating as intended [9].

Conclusion

The continuous process improvement in software testing not only ensures higher product quality but also optimizes business processes. Software quality is the degree to which the correct software produced. Quality software is reasonably bug or defects free, delivered on time and within budget, meets requirements, expectations, and maintainable. Software life cycle testing essentially means that testing occurs in parallel with the development cycle and is a continuous process. It is important to start the software testing process early in the application life cycle, and it should be integrated into application development itself. Quality control and quality assurance play a key role in improving the quality of software.

References

1. Imam A, Khoja S. A, Shariff I (2007) Improving software quality—a benchmarking approach. Conference Paper
2. Hegedus P, University of Szeged, Hungary (2015) Advances in software product quality. Measurement and its applications in software evolution, ICSME 2015. IEEE, Bremen, German, pp 590–593
3. Software Quality Professional (2006) "Are standards the answer?" ASQ Publication 8(2):p 8
4. https://testpoint.com.au/11-ways-to-improve-software-quality/. Accessed 16 Aug 2022
5. https://towardsdatascience.com/14-ways-to-improve-software-quality-other-than-testing-a62d14936575. Accessed 16 Aug 2022
6. https://qxperts.io/5-ways-to-improve-software-quality. Accessed 16 Aug 2022
7. https://marutitech.com/software-testing-improvement-ideas/. Accessed 16 Aug 2022
8. https://www.altexsoft.com/blog/engineering/software-testing-qa-best-practices/. Accessed 16 Aug 2022
9. https://www.softwaretestinghelp.com/improve-software-quality/. Accessed 16 Aug 2022

Part IV
Validation

Chapter 9
General Principles of Validation

Abstract You can enhance the quality of your work through validation. It is a crucial step in the process of constructing quality because it employs a preventative approach. The facilities, systems, tools, manufacturing processes, software, and testing methods that have an impact on a product's quality, safety, and effectiveness are verified through validation, a scientific quality assurance examination. When utilizing new products that generate outcomes that are similar to those of older ones, validation has become an important research area. Today, validation is employed in a wide range of disciplines, including science and medicine. Additionally, the validation procedure is closely linked to technological advancement (software, computers, tools to collect the data needed). The three stages of validation activities that can all be carried out using statistical process control are pre-validation, process validation, and validation maintenance. This chapter covers the advantages, components, need, scope, purpose, and principles of validation. The duties of numerous departments during the validation process are also covered in this chapter.

Keywords Validation · Quality assurance · Confirmation · Effectiveness · Quality

Introduction

"Validating validation actions or establishing effectiveness" is what the phrase "validation" means. Validation is the documented act of proving that any process or activity consistently produces the desired results. It also involves the certification of apparatus and systems. Validation does not make processes better; rather, it verifies that they have been correctly configured and are operating as intended. Validation should be planned to guarantee product quality, regulatory compliance, and safety while ensuring consistency and reliability. Technical equipment qualification and process validation are both increasingly referred to as "validation" in the same sentence. A method, operation, activity, material, system, or piece of equipment's ability to provide the intended results is determined through the validation process [1].

S. S. Shukla et al., *Pharmaceutical Calibration, Validation and Qualification: A Comprehensive Approach*, https://doi.org/10.1007/978-981-19-9002-1_9

Benefits/Importance of Validation

- Quality.
- Customer—patient satisfaction.
- It has been built into the product.
- Assurance of quality.
- Time bound.
- Process optimization.
- Reduction of quality cost.
- Nominal mix-ups, and bottle necks.
- Minimal batch failures leads to improved efficiency and productivity.
- Reduction in rejections.
- Increased output.
- Avoidance of capital expenditures.
- Fewer complaints about process-related failures.
- Reduced testing in process and in finished goods.
- More rapid and reliable start-up of new equipment.
- Easier scale-up forms development work.
- Easier maintenance of equipment.
- Improved employee awareness of processes.
- More rapid automation.
- Government regulation.
- Validation requirements is necessary for obtaining approval to manufacture and to introduce new products) [1].

Definitions of Validation

Validation is defined as per various bodies EC, ICH, FDA, WHO as given below:

According to European Commission

1991: Validation-"Act of proving, in accordance of GMPs that Any..." process actually leads to expected results.

2000: "Documented evidence that the process, operated within established Parameters, can perform effectively and reproducibly to produce a Medicinal product meeting its redetermined specifications and quality attributes" [2].

According to US FDA Definition

"Process validation is establishing documented evidence which provides a high degree of assurance that a specified process will consistently produce a product meeting its pre-determined specifications and quality characteristics" [2].

According to ICH Definition

"Process validation is the means of ensuring and providing documentary evidence that processes within their specified design parameters are capable of repeatedly and reliably producing a finished product of the required quality" [2].

According to WHO Definition

"The documented act of proving that any procedure, process, equipment, material, activity or system actually leads to expected result".

Ted Byers and Bud Loftus, two FDA (Food and Drug Administration) officials, suggested the idea of validation to raise the calibre of pharmaceutical products in the middle of the 1970s. In response to an increase in problems with parental sterility products, it was proposed. When these pharmaceuticals were originally being manufactured, the procedures were the focus of validation efforts, but over time they expanded to cover all pharmaceutical manufacturing processes [3].

Need/Purpose of Validation

In order to assess the system's quality at each stage rather than simply at the end, validation methods include inspections of manufacturing materials, operational procedures, personnel training, and system monitoring during production. The concept of validation has expanded in recent years to include a variety of activities, from quality control techniques for medications and medical substances to computerized systems. An essential part of CGMPs today is the validation process. A new or improved method must be validated in order to determine its precision, accuracy, and detection limit, limit of determination, selectivity, linearity field, and method transferability [3].

Numerous disciplines, such as business, economics, psychology, chemistry, and biology, can all benefit from validation. The following are the three main arguments in favour of the need for validation:

- A prerequisite for the complete system for ensuring the quality of products.
- A person who is capable of attaining the desired outcome.
- Assures that each batch of any product released onto the market continuously complies with all quality standards [4].

Validation Is Needed

- Whenever the circumstances under which a methodology has been validated change, for as when an instrument's characteristics change, or before regularly implementing a new approach.
- The initial use of a new substance or medication [4].

Purpose of Validation

- Using one sample to represent the whole population being studied.
- Enabling sample participation in the measurement procedure.
- To reduce the number of ensuing queries about the reliability of the sample.
- Including a provision for re-sampling, if necessary [4].

Scope of Validation

- Show the flexibility to learn a new production method or recipe.
- A report in writing on the results will be created.
- Long-term validation, such as at least three successive batches of the entire manufacturing size, to demonstrate consistency (worst-case scenarios should be taken into account) [4].

Since pharmaceutical validation is such a broad subject that it essentially encompasses every aspect of pharmaceutical processing, defining the scope of validation is a challenging undertaking. However, a comprehensive analysis of pharmaceutical activities will reveal at least the following areas for pharmaceutical validation [4].

- analytical
- instrument calibration
- process utility services
- raw materials
- packaging materials
- equipment
- facilities
- manufacturing operations
- product design
- cleaning
- operators

Reasons of validation

- customer satisfaction
- product liability
- lower manufacturing costs
- aids in process improvement [4]

Principles of Validation

Quality, safety, and efficacy can be summed up as the three validation principles in product design and construction. Inspection of the finished product and work-in-progress are insufficient for quality assurance. The production process is reviewed at each stage to ensure that the final product meets all requirements for quality. Current Good Manufacturing Practices (cGMP) for completed pharmaceuticals, which are outlined in 21 CFR sections 210 and 211 of the FDA rules, serve as a broad reference for validation. To ensure that in-process components and the end product fulfil specified quality criteria consistently and reliably, manufacturing processes must be designed and regulated in accordance with cGMP regulations. GMPs are seen to be incomplete without validation. Global validation requirements must be met in order to obtain authorization to create and introduce new products [4].

Documents Used in Validation

1. validation master plan
2. validation protocol
3. validation report
4. SOPs.

Validation Master Plan

The three main validation principles in product design and construction are quality, safety, and efficacy. Inspection of the finished product and work in progress alone cannot ensure quality. To ensure that the final product meets all quality requirements, every step of the production process is examined. FDA standards found in 21 CFR parts 210 and 211, which provide current good manufacturing practices (cGMP) for finished pharmaceuticals, serve as a broad reference for validation. In order for in-process components and the end product to consistently and reliably fulfill stated quality requirements, manufacturing processes must be designed and controlled in accordance with cGMP regulations. Validation is thought to be a crucial component of GMPs. Validation requirements must be completed globally in order to obtain approval to create and introduce new products [5].

The facilities, systems, tools, and practices that will be validated are listed below:

- Validation procedure.
- Organizational structure for validation activities.
- The documentation format is the format in which protocols and reports will be produced.
- It is impossible to emphasize the value of scheduling and planning.
- Make a control modification.
- There is a mention of previously released content.

- It could be essential to construct numerous validation master plans for large projects [5].

Validation Protocol

A validation protocol is a set of guidelines that specifies the steps that will be taken to validate a process, the people who will be in charge of each task, the testing parameters, sample plans, testing methodology, and the requirements that will be used. Additionally, it describes the features of the product and the tools that will be employed. The minimum number of batches required for validation studies must be specified, together with the acceptance standards and the individuals who will sign, approve, or reject the study's conclusions. Form and process are the two sections of the validation protocol. Standard operating procedures (SOPs) outline precisely what has to be confirmed [5]. The validation protocol must be numbered, signed, and dated, and must at least contain the following details:

- Title
- Objective and Scope
- Responsibility
- Protocol Approval
- Validation Team
- Product Composition
- Process Flow Chart.
- Manufacturing Process
- Review of Equipment/Utilities
- Review of Raw Materials and Packing Materials Review of Analytical and Batch Manufacturing Records
- Review of Batch Quantities for Validation (Raw Materials)
- Review of Batch Quantities for Validation (Packing Materials)
- HSE Requirements
- Review of Process Parameters Validation Procedure
- Sampling Location.
- Documentation
- Acceptance Criteria
- Summary
- Conclusion

Validation Protocols consist of mainly:

- A description of the sampling procedure, including the kind and quantity of samples taken; a description of the validated process, apparatus, or method.
- Standards for accepting test results.
- Revalidation standards or a timetable [6].

Validation Summary Report

To confirm that procedures have been validated, the Validation Support Record (VSR), a legally required document, lists all current validation documentation and details any revalidation requirements. On the Process Validation Master Plan, each VSR document number is listed for quick retrieval. The validation report's common format includes the following:

- Executive Summary
- Discussion
- Conclusions and Recommendation
- List of Attachments

The aforementioned issues should be brought up in the protocol's sequence of appearance, with any deviations being justified. The representatives of each unit must sign and date the report [6].

SOP (Standard Operating Procedure)
The general format of the SOPs involves:

- Title
- Code
- Objective
- Scope
- Definitions
- Description
- Safety
- Documentation
- Effective date, review date, version number
- Footer: Prepared By, Reviewed By, Approved By, Authorized By
- References

Validation Setup

The desirable characteristics, which encompass both chemical and physical properties, are chosen during validation setup. For parenterals, stability, the absence of pyrogens, and the absence of discernible particles are all desirable qualities. To ensure uniformity and consistency in the desired product attributes, product acceptance specifications should be created. They should result from testing and challenging the system on a sound statistical basis both during the initial development and production phases and throughout ongoing routine production [6].

The technique and tools should be chosen to satisfy the product's requirements. Design engineers, production workers, and quality assurance staff may all be involved. The method should be painstakingly comprehensive, and every step should be rigorously verified to confirm its efficacy. These factors are essential for ensuring consistent levels of product quality, purity, and performance [6].

Roles and Responsibilities of Validation Team (Table 9.1)

Table 9.1

Department	Designation	Responsibility
Research and Development (R&D)	Executive/ officer	To coordinate the entire validation process, it is necessary to schedule meetings and interactions with production, quality control, and quality assurance. The creation of a preliminary validation protocol, a master formula record, the monitoring of the process, the compiling and analysis of the data, and the final report. To go over the initial validation documentation
Quality assurance	Officer	To arrange team meetings and discussions in order to plan the entire validation procedure. Creating the validation technique, monitoring the process, compiling and analysing the data and test results, and creating the final report. Review of the validation documents is required
Production	Officer	To aid in the production process's validation procedure. To help with the data gathering procedure
Quality control	Officer	Testing, then reporting test results
Quality assurance	General manager	To accept the process's validation protocol and report. Review of the validation documents is required

Validation Life Cycle

Validation is a dynamic and ever-evolving process. The validation process comprises an in-depth theoretical analysis of the system's and processes' operation. This analysis can range from the most fundamental to the most complex. Its scope includes document revision control, training, and system and process upkeep.

All throughout the organization, the management structure should be transparent and show validation proof.

Setting and upholding strict standards is done through validation.

The three stages of validation activities that can all be carried out using statistical process control are pre-validation, process validation, and validation maintenance [6].

Conclusion

Validation is an important step in establishing and maintaining the quality of the final product. If each step of the production process is examined, we can ensure that the end product is of the highest quality. Validation is the skill of generating and practicing the processes as well as the documentation that have been designed. In the disciplines of medicine research, manufacturing, and finished product specification, the term "validation" is most typically employed. Consistency and reliability of a proven process to offer a quality product are crucial for an industry.

References

1. Ahir KB, Singh KD, Yadav SP, Patel HS, Poyahari CB (2014) Overview of validation and basic concepts of process validation. Sch Acad J Pharm 3:178–190
2. Lendvai AMG (2018) Validation—a brief introduction. Geo Patterns 3(1):10–15
3. Agalloco J (2011) Validation: an unconventional review and reinvention. PDA J Pharm Sci 49(4):175–179
4. Sharma S, Khurana G, Gupta R (2013) A review on pharmaceutical validation and its implications. Indian j pharm biol res 1(3):100–104
5. Murthy DN, Chitra K (2013) Review article on process validation as a quality assurance tool a theoretical approach. Int J Pharm Sci Res 3(2):6–14
6. Haider SI (2002) Validation standard operating procedures: a step by step guide for achieving compliance in the pharmaceutical, medical device, and biotech industries. St. Lucie Press

Chapter 10
Process Validation and Its Types

Abstract Validation for pharmaceuticals ensures that the production procedure is dependable and repeatable. Effective process validation is essential for assuring drug quality. The fundamental tenet of quality assurance is that a medicine should be produced in a way that makes it appropriate for the use for which it is intended. Process validation is defined as the collection and analysis of data that provide scientific proof that a process is capable of reliably producing high-quality output from the process design stage to commercial production. A number of tasks are involved in process validation, which happens throughout the lifecycles of both products and processes. Process Design, Process Qualification/Validation, and Continued Process Verification are the three phases of process validation. We have talked about the regulatory requirements for process validation from the FDA, WHO, GMP, and ISO.

Keywords Prospective · Revalidation · Process validation · Design · Improvement

Introduction

Definition of the Term "Validation"

Validation is defined as "the assurance, through the provision of objective evidence, that the requirements for a certain intended use or application have been met" (ISO 9001:2015).

According to the standard, the objective evidence required for validation is the outcome of a test or another sort of determination, such as alternative computations [1].

Definition of the Term "Process"

When it comes to the definition of the term "process", ISO 13485 refers to ISO 9000: 2015 once again. A process is defined as a "collection of interrelated or interacting

S. S. Shukla et al., *Pharmaceutical Calibration, Validation and Qualification: A Comprehensive Approach*, https://doi.org/10.1007/978-981-19-9002-1_10

activities that employ inputs to accomplish an intended outcome" according to this standard. Example processes are [1]:

- development process
- sterilization process
- production process
- recruitment process
- sales process

Process Validation

Clearly, process validation is the objective, evidence-based affirmation of a process that yields the anticipated process outputs when the two criteria are combined.

For instance, in a development process, one would make sure that the results meet the requirements ("Design Input"). To ensure that an item is truly sterile before sterilizing it, one would check its sterility [1].

The FDA defines "process validation" as "the establishment through objective evidence that a process consistently produces a result or product that satisfies its intended requirements" [1].

Definition of Process Validation

To establish scientific proof that a process is able to reliably produce high-quality items, process validation is the collection and analysis of data from the process design stage through production [1].

Process validation is required by current Good Manufacturing Practices (GMPs) for completed pharmaceuticals (21 CFR 211) and GMP regulations for medical devices (21 CFR 820), and as such, it is applicable to the manufacturing of both drugs and medical devices.

The lifecycle of a product or process includes a number of tasks that make up process validation [1].

The following definition is one of the standards for process validation that the US Food and Drug Administration (FDA) has published: The production of documented proof that a certain process consistently produces a product that satisfies its predetermined characteristics and quality criteria is known as "process validation" [2].

To ensure that a process is consistently produced in accordance with the established standard, it is necessary to examine data collected throughout product design and manufacturing [2].

Validating processes aims to provide consistent high-quality output. Guidelines on process validation have been released by regulatory organisations including the

Food and Drug Administration (FDA) and the European Medicines Agency (EMA) [2].

The FDA claims that the selection of high-quality components and materials, appropriate product and process design, and process (statistical) control through in-process and end-product testing are all crucial variables that contribute to product quality assurance. Thus, by careful design (qualification) and validation of both the process and its control systems, a high level of confidence may be developed that each individual manufactured item in a specific batch or succession of batches that meets criteria will be acceptable [2].

Process validation promotes modern manufacturing concept including, among other things, such as process improvement, innovation, good science, risk assessment, and mitigation.

Importance of Process Validation

- The procedure is optimised, and product quality is guaranteed.
- The cost of a high-quality product is reduced.
- Recalls of products from the market are reduced.
- The procedure is under control; thus, it is possible to conduct a complete investigation.

Process validation program can be made more effective and efficient through

- more effective project management
- scientific knowledge administration and preservation
- techniques for acquiring and analysing data
- applying a multidisciplinary perspective
- properly documented project plans
- senior management involvement statistical analysis of data [3]

The process validation activities can be described in three stages

- Stage 1—Process Design
- Stage 2—Process Qualification/validation
- Stage 3—Continued Process Verification

Stage 1—Process Design: The commercial process is established at this phase based on the knowledge amassed throughout development and scale-up activities. In this phase of research and development, the product's manufacturing process is defined. Typically, it includes the following things:

- making a Quality Target Product Profile (QTPP)
- defining Critical Process Parameters (CPPs)
- identifying Critical Quality Attributes and performing risk assessments

Conducting the Process Design involves

- Research can take many different forms, including retrospective analysis, process risk assessment, experiment design, modelling (which forecasts and validates process/parameter behaviour), and laboratory, engineering, pilot, small-scale and commercial scale studies.
- Develop a Containment Plan—For both the overall process and each unit operation, divide the process into segments. Concerns about the plan include operational limitations and legal restrictions, as well as commercial production and control records.
- The Control Strategy should be sent to the next step for verification (PQ).
- Verification or determination of the restrictions on the crucial process parameters. Boundaries for the procedure and its limits have been set. Learning about process factors makes it possible to create a control strategy [3].

Stage 2—Process Qualification: At this point, the process design has been verified as being capable of reproducible commercial manufacture, which includes qualification of the facility, utilities, and equipment. This stage needs to be properly finished before commercial release. The process qualification steps include the following:

- design of the facility and qualification of equipment
- performance qualification
- PPQ protocol

All operations throughout the Process Qualification must adhere to cGMP guidelines. A deep grasp of the product and process should serve as the basis for PQ. The PQ lots should be created under usual circumstances by individuals who are anticipated to complete each task regularly. A successful PQ will validate the design of the process and demonstrate that the commercial production process ran according to schedule. To ensure that the earlier-designed process can produce reliable and consistent outcomes, this stage evaluates and qualifies it [3].

It involves compiling and examining data pertaining to all facets and stages of the production process. This includes:

- The building and facilities, including ensuring that they abide by local and legal regulations governing pharmaceutical manufacture.
- Transporting raw resources to their intended destinations.
- Storage for raw materials.
- The knowledge, training, and work practices of those working in the production line.
- Every stage in the conversion of raw resources into finished products. This includes predetermined sample sites at different stages of the procedure.
- Final-goods packaging, storage, and delivery.

Stage 3—Continued Process Verification: It includes upkeep, continual verification, and process improvement. Continuous assurance that the process is in control is obtained during ordinary production. Data collection and monitoring are done when the commercialization process is being examined. This phase's goal is to guarantee that the process stays in check during commercial production. After the equipment qualifying state has been attained by regular monitoring, maintenance, and calibration procedures, it must be maintained. Additionally, the data should be regularly assessed to see whether requalification is required. It is necessary to create a constant programme for gathering and analysing data about products and processes. The data acquired should demonstrate that the crucial quality characteristics are upheld throughout the process, final product packaging, storage, and delivery [3].

On-going programme sampling is present. At the initial stage of commercial production, monitoring and/or sampling must continue at the level decided upon during the process qualification stage. The second stage of commercial production starts once there are sufficient data points to generate reliable estimates of the variability and once the variability is known. It is important to modify sampling and/or monitoring to a statistically appropriate and representative level [3].

Process Optimization

The data gathered in Stage 3 might provide ways to enhance and/or optimize the process. To determine whether any adjustments to the current procedure are required, information and data on product performance and production experience should be checked on a regular basis. The improvements may necessitate additional process design and process qualification activities [4].

Ongoing the goal of continuous process verification is to identify process drift by gathering and analysing data and information on the process' performance. Process maintenance includes a critical component called ongoing product performance feedback [4].

Product quality is used to describe a product's consistency in performance from batch to batch and unit to unit in the context of process validation [4].

Continuous Process Verification involves ongoing testing while commercial products are being produced to make sure the previously designed and validated process still provides reliable quality [4].

Types of Process Validation

Depending on when they are carried out in connection to production, different types of process validation are used; the recommendations on general principles of process validation include four types of validation:

- prospective validation (or premarket validation)
- concurrent validation
- retrospective validation
- revalidation [4]

Prospective Validation (or Premarket Validation)

Establishing in writing, prior to process implementation, that a system performs as promised based on pre-planned protocols. If a new recipe (or a new facility) needs to be verified before routine pharmaceutical manufacture can start, this method of validation is frequently used. The manufacturing process is typically moved from development to production as a result of employing this approach to validate a process. Prospective validation is done at the development stage (also called premarket validation). The outcomes of prospective validation assist in evaluating the manufacturing process's risk analysis. It is divided into steps and assesses crucial elements such as temperature, mixing time, and relative humidity during the production process.

- To launch new products in the manufacturing plant.
- This validation shall be taken into consideration for the first three successive production size batches after process stability if there is a significant change in the manufacturing process and the impact of the modifications, for example, a leak test that failed because the blister's sealing was insufficient [5, 6].

Criteria for Prospective Validation

- Three potential validation batches should all have the same settings.
- These three batches must all have the same procedure parameters.
- The first batch will be dispatched following the production, testing, and evaluation of the third batch [5, 6].

If some production process parameters need to be changed after the first batch, follow these guidelines.

- The first batch should not be included in this validation. Instead, create three batches with the modified parameters.
- Consider the following batches as potential validation batches with the same requirements: 2, 3, and 4.
- A suitable change management system must be used when putting a change into action.
- The first validation batch will be made available for purchase and distribution following the production, testing, and evaluation of all three batches.
- The results from all three batches need to be satisfactory [5, 6].

Concurrent Validation

Based on data gathered while the process is really being carried out, concurrent validation is a way for establishing documented evidence that a facility and its processes actually carry out what they claim to. This strategy entails keeping an eye on crucial processing steps and testing the finished product to show that the manufacturing process is under control. Concurrent validation is done during the routine manufacturing step. It is founded on an exhaustive evaluation of the procedure, which is supported by upcoming validation. The initial three batches of large-scale production must be properly watched [5, 6].

When Concurrent Validation is Carried Out?

A manufacturing facility has prospectively verified a new product; the manufacturing process is unaffected, and the effects of the adjustments are negligible. The formula also changes if the current production's raw material source does. Examples of in-process testing include pH value, tablet hardness, weight variation, dissolution time, uniformity of content, viscosity or density, uniformity of colour, particle size distribution, average unit potency, and others [5, 6].

Retrospective Validation

Retrospective validation is applied to operational facilities, processes, and process controls that have not gone through a well-established validation procedure. These facilities, procedures, and process controls can be validated using historical data, which provides the necessary documentary proof that the process is operating as intended. Because of this, this kind of validation is only appropriate for established processes and is useless when the equipment, operating procedures, or product composition have recently changed.

This tactic is rarely used today because it is quite unlikely that any existing products haven't been submitted to the prospective validation procedure. It is only used to audit a process that has been validated [5, 6].

Retrospective validation uses historical and testing data from prior produced batches. It includes a trend analysis and the degree to which the process falls within the permissible range of the process parameters [5, 6].

The next method for performing validation makes use of either computer-based data or a manual approach (retrospective validation).

- Organize all data sequences in a logical order, such as batch production and expiration date. Obtain information from previously finished batches.
- Ten to twenty-five batches, or more, are employed for this purpose, and they are preferably processed over a period of no more than 12 months and reviewed concurrently [5, 6].

Revalidation

Revalidation comprises a review of past performance data as well as a partial or complete repetition of the initial validation effort. Maintaining the validated status of the production processes, computer systems, and plant equipment requires this technique. Revalidation is necessary to make sure that modifications to the process environments, whether intentional or unintentional, do not adversely affect the process characteristics and product quality [7].

The possible reasons for starting the revalidation process include:

- Transferring a product from one factory to another; alterations to the item, the facility, the production process, the cleaning procedure, or other elements that might have an effect on the product's quality.
- The significance of routinely reviewing the validation results.
- A material modification to the batch size (typically by an order of magnitude).
- Sequential batches of products and processes that don't fulfil criteria.
- The size of the adjustments and their effect on the product determine the breadth of the revalidation procedures [7].

Revalidation is needed when:

- following any change that has an impact on product quality
- revalidation on a regular basis
- reduction in batch size
- facility and plant changes [7]

Regulatory Requirements for Process Validation

Requirements of ISO 13485

- Transferring a product from one factory to another; alterations to the item, the facility, the production process, the cleaning procedure, or other elements that might have an effect on the product's quality.
- The significance of routinely reviewing the validation results.
- A material modification to the batch size (typically by an order of magnitude).
- Sequential batches of products and processes that don't fulfil criteria.
- The size of the adjustments and their effect on the product determine the breadth of the revalidation procedures [7].

If those conditions are met, process validation is required. For this purpose, the company must determine procedures including the following:

- process evaluation criteria.
- people qualification.

- method and procedure application.
- records and documentation.
- revalidation.
- approval of process modifications [7].

Further Relevant National and International Provisions

The Lander ZLG's Central Authority has issued a document titled "Validation of Production and Providing Services (including Software)" with the number ZLG 3.9 B 18 [7].

Requirements of the FDA

In the Quality System Regulations, specifically in 21 CFR 820:75: The FDA specifies the requirements for process validation in "Process Validation". Validation is necessary if the process results cannot bc independently verified. These validation initiatives need to incorporate the following:

- All completed tasks must be recorded, together with the date and signer.
- Procedures must be established for observing proccss parameters.
- Only trained individuals are capable of validating a procedure.
- Methods and information used to regulate and monitor processes, as well as the date of execution, the people who will be performing the validation, and the necessary tools, must all be documented.
- The maker is responsible for determining whether revalidation is necessary in the case of changes and, if so, for carrying it out.
- In its "Guidance Document Software Validation", the FDA outlines the criteria for validating software.

For industry leaders to ensure consistency in data collection and other information gathering processes as well as to improve accessibility so that benefits can be realized later in the product lifecycle, the FDA makes a number of recommendations. The FDA promotes a team-based method of process validation, where professionals from many fields work together to develop best practices and procedures. There is a lot of potential in numerous studies that aim to explore, uncover, and observe data regarding the pharmaceutical product and technique. These are strongly recommended, provided that all study protocol requirements are met. All parameters (process, operational, and equipment) and attributes (quality, product, and component) should have controls that are commensurate to the level of importance of each [8].

Requirements of the WHO

Current good manufacturing practices (cGMP) are crucial for validating pharmaceutical process. Many of them have legal duties attached to them [8].

The FDA has determined that a finished drug product will become adulterated (i.e. will become lower in quality by adding another chemical) when requirements cannot consistently and reliably attain a pre-specified quality in manufacture, processing, packing, or holding [8].

To confirm any production processes that might be causing variability, certain control procedures are also needed for sampling and testing. All samples must conform to the batch's established specifications, be representative of the batch under examination, and fulfil statistical confidence requirements.

Data on product quality and production processes must be assessed and adjusted on a regular basis in accordance with Sect. 211.180(e). Therefore, the idea of continuous review is essential to cGMP compliance in order to gather ongoing feedback on good manufacturing practices [8].

Finally, cGMP mandates that the size, construction, and placement of drug production facilities and equipment be sufficient to meet the necessary standards. According to industry requirements, every equipment must be inspected and calibrated [8].

To ensure that products are what they seem to be in terms of identification, strength, quality, purity, and potency, regulators naturally require written protocols for production and process control [9].

Order of Priority in Process Validation

List the many product categories that need to be confirmed because it is hard to check a company's entire offering. Priority is given to the company's most lucrative products [9]. Regarding the importance of product and process validation, the following is suggested:

1. *Sterile Products and Their Processes.*
   (a) Large-Volume Parenterals (LVPs).
   (b) Small-Volume Parenterals (SVPs).
   (c) Ophthalmic, other sterile products and medical devices.
2. *Non-sterile Products and Their Processes.*
   (a) low-dose/high-potency tablets and capsules/transdermal delivery systems (TDDs)
   (b) drugs with stability problems
   (c) other tablets and capsules
   (d) oral liquids, topicals, and diagnostic aid

Documentation

Documentation at each level of the process validation lifecycle is essential for efficient communication in challenging, protracted, and heterogeneous projects. To guarantee that information learned about a process or product is available to and comprehended by everyone involved at every point of the lifecycle, documentation is essential. The scientific approach is based on openness and accessibility of knowledge. A product's commercial release ultimately depends on the responsible and accountable organizational units involved in the process being able to make informed, science-based decisions. Diverse levels and types of documentation are required by CGMP at different stages of the validation lifecycle. The greatest paperwork is needed during Stage 2, process qualification, and Stage 3, continuing process verification. These phases must be conducted in accordance with CGMPs, and research must receive the quality unit's approval [9].

Process Validation of Various Dosage Forms

The production, testing, packaging, and marketing of these goods all fall under the purview of the process validation of dosage forms (solids, semisolids, liquids, and parenterals). All API, drug, excipient, formulation, and process variables must be monitored and managed at every stage for optimal batch and final product optimization. The chapters that follow discuss the process validation of solid, semisolid, and liquid dosage forms [9].

Manufacturing Process

The standard production processes for pharmaceuticals typically involve the use of equipment, protocols, people, materials, measuring systems, and environmental factors, among other things.

The following are crucial factors the maker should take into account:

- Understanding process variations, detecting and evaluating them, understanding the impact on the process and the product, and controlling such variations depending on the risk they pose are all important.
- Formal testing protocols for each batch must be developed and followed in order to guarantee batch uniformity and integrity of pharmaceutical products.
- Such control techniques are designed to keep track of results and verify the effectiveness of manufacturing procedures that may be the cause of process variability, as measured by statistical methods, in in-process material and medicinal product qualities [9].

Process Control

There are several contemporary techniques for process control and validation, including:

- Six Sigma
- Statistical Process Control
- The Process Capability Index

Examples of statistical process control include a sampling strategy, experimental design, variance reduction, process capability analysis, and process improvement strategies (SPC). SPC can be used to maintain consistency in the production process, but it won't make a poorly designed product more reliable. Setting up in-process requirements: The past acceptable process average and process variability as determined by statistical procedures are used to specify process parameters whenever possible [9].

Responsibilities of Process Validation Team Product [9] (Fig 10.1)

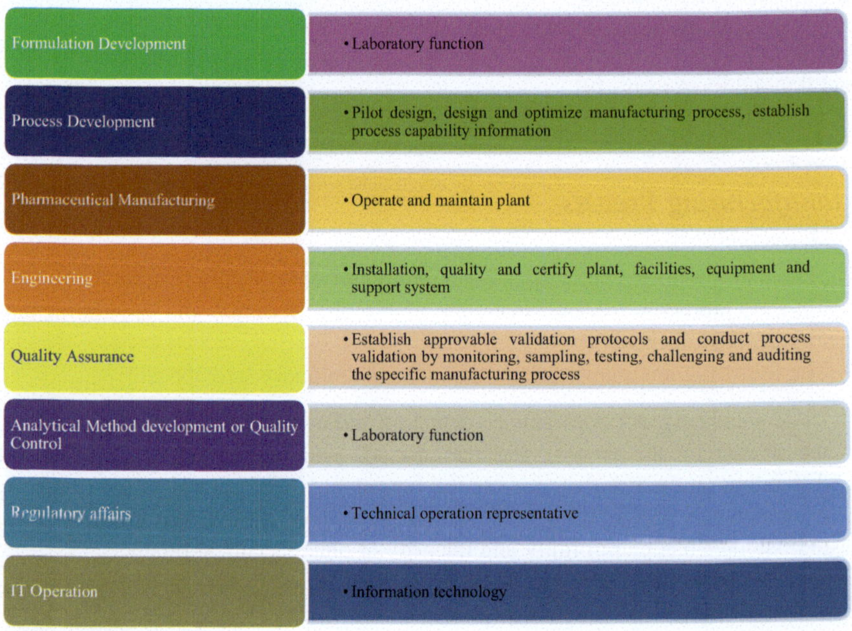

Fig. 10.1 Responsibilities of process validation team product

Conclusion

Process validation assures product quality, uniformity, and reproducibility, as well as the safety of pharmaceutical products, as mandated by regulatory bodies all over the world. To ensure that the product meets all quality, manufacturing, and regulatory criteria, the multidisciplinary validation team must first identify, evaluate, and incorporate the crucial needed validation key parameters. The most significant and well-known cGMP parameter is process validation. The process validation is meant to help producers understand the needs of their quality management system (QMS) for process validation, and it may be applied to any manufacturing process.

References

1. Kaur H, Singh G, Seth N (2013) Pharmaceutical process validation: a review. J Drug Deliv Ther 3(4):189–194
2. SFDA (2010) Guidelines for process validation of pharmaceutical dosage forms. Saudi Food and Drug Authority, Dammam, Saudi Arabia, pp 9–15
3. Mohammed SA (2012) Pharmaceutical process validation: an overview. J Adv Pharm Res 2(4): 186–200
4. Ahir KB, Singh KD, Yadav SP, Patel HS (2014) Overview of validation and basic concepts of process validation. Sch Acad J Pharm 3(2):178–190
5. Sarvani V, Elisha RP, Nama S, Pola LM, Rao CB (2013) Process validation: an essential process in pharmaceutical industry. Int J Pharm Chem Anal 3(2):49–52
6. Guidance for Industry: Process Validation: General Principles and Practices (2011). U.-S. Department of Health and Human Services, Food and Drug Administration, Centre for Drug Evaluation and Research (CDER), Centre for Biologics Evaluation and Research (CBER), Centre for Veterinary Medicine (CVM)
7. Nash RA, Wachter AH (2003) Pharmaceutical process validation: an international, 3rd edn. Revised and Expanded, Marcel Dekkar, Inc., New York, pp 17–40
8. Kathiresan K, Moorthi C, Prathyusha Y, Gade BR, Reddy BK, Manavalan R (2010) An overview of pharmaceutical validation. Res J Pharm Biol Chem Sci 1(4):1026
9. Nandhakumar L, Dharmamoorthy G, Rameshkumar S, Chandrasekaran S (2011) An overview of pharmaceutical validation: quality assurance view point. IJRPC 1(4):1003–1014

Part V
Analytical Validation

Chapter 11
Pharmaceutical Equipment Validation

Abstract The process of approving documentary proof that a following procedure, method, or action will consistently provide the desired result is known as validation (predetermined requirements). The validation programme in the pharmaceutical sector includes process validation, cleanliness, facilities, equipment, and instruments. In this review post, we will provide a brief explanation of equipment validation, one of the most widely used methods of validation. We will discuss the many forms of equipment validation, the paperwork needed, and the uses and importance of validation in the pharmaceutical industry. Before new equipment or instruments may be verified, pharmaceutical enterprises must now comply with a regulatory obligation known as equipment validation. While validation calls for in-depth knowledge about the instrument being validated, it is normally carried out by the company that supplies the equipment.

Keywords Validation · Documentary · Evidence · Equipment · Pharmaceutical industries

Introduction

The validation process is the written justification that ensures the planned conformance and desired outcome. The term validation is often used in the pharmaceutical sector. The Food and Drug Administration (FDA) first created the validation idea in the mid-1970s to improve the quality of pharmaceutical products. The phrase "valid or validity" refers to something that is "legally defined". Due to the fact that a variety of practices, techniques, and activities are validated to guarantee their high quality [1–3].

Types of Validation

Validation is divided into following subsections which include: [3, 4].

S. S. Shukla et al., *Pharmaceutical Calibration, Validation and Qualification: A Comprehensive Approach*, https://doi.org/10.1007/978-981-19-9002-1_11

- analytical method validation
- process validation
- cleaning validation
- equipment validation

Equipment Validation

Equipment validation is a technique that is recorded to ensure that any piece of equipment works as intended and generates findings that are acceptable and correct (predetermined result). The premise that equipment must be developed, produced, maintained, and modified in order to carry out the necessary operations forms the basis of the equipment validation process. Before beginning any activity, equipment validation is essential because it is the fundamental element of the pharmaceutical sector (documented evidence of equipment) [5, 6].

Types of Equipment Validation

The process of equipment validation is not a one-step activity; it is divided into stages, each of which has its own subsections or processes, as follows: [6].

- design qualification
- installation qualification
- operational qualification
- performance qualification
- process qualification

In the pharmaceutical industry, certification of pharmaceutical equipment is a simple procedure. In accordance with the pharma industry's/clearance, several steps of the process are thoroughly investigated and documented. The initial phase in the procurement process is typically the creation of the necessary papers and user requirement specifications (URS). To finish the validation project/plan, a change request (CR) type should be taken from the existing facilities (VP). A request to carry out a validation project has been made, as previously authorized by management (VP). Once the VP has been approved, the validation protocol can make sure that all of the conditions outlined in the URS and cGMP guidelines are as follows.

Phases of Equipment Validation

The process of equipment validation is divided into three phases: [6, 7]

- Phase 1: Pre-validation phase
- Phase—2: Process validation phase
- Phase—3: Validation maintenance phase

Pre-validation Phase

Design Qualification (DQ)

It is a documented assessment of the design of the machinery and industrial facilities. Design qualification's main objective is to make sure that all system requirements are correctly established from the beginning. The design qualification process will demonstrate that all quality concerns are considered throughout the design phase. It details all requirements stated in the user requirement specification (URS), as well as any cGMP laws and regulations, as well as the instrument's functional and operational specifications. The offering of the design must be qualified in writing to guarantee that it will follow the given rules: [8].

- Functional specification (FS).
- Tender specification and drawing.
- Purchase specification.
- Vendor qualification.
- User requirement specification (URS): It includes the list of requirements/expectations of the customer in the equipment. The general customer requirements are as follows:
- Size of equipment and space occupied by it.
- Effectiveness and durability of the equipment.
- Working speed of the equipment.
- Equipment should be with low noise and air pollution.
- Availability of the spare parts and also provides services at minimal cost.
- Overall good construction.

Installation Qualifications (IQ)

The installation qualification confirms that the specified equipment, including all parts, spares, service gauges, and other necessary materials, was delivered and installed in accordance with the planned schedule and in flawless condition. It serves as written proof that the apparatus was correctly fitted and calibrated. The purpose of IQ is to ensure that all facets of the equipment are set up correctly and adhere to the original (URS) design. According to the manufacturer's installation guidelines, the working conditions have been documented and verified as being suitable for the instrument's operation [9].

The documentation of installation includes:

- details of supplier and manufacture
- equipment name, colour, model, and serial number
- date of installation and calibration

Process Validation Phase

Operational Qualifications

The installed equipment/instrument will function flawlessly under the designated operating conditions, thanks to operational certification. Additionally, it guarantees that the testing results are accurately recorded and that the equipment is in perfect functioning condition to meet predetermined performance parameters. The objective of operational qualification is to confirm that all dynamic conditions are consistent with (URS) design. It includes criteria for verification and traceable electric stimulators to make sure the machinery is operating as it should. Operational qualification gave users a high level of trust that the equipment complies with both their needs and the manufacturer's criteria in terms of functionality (URS). Operational qualification, also known as process validation, ensures that equipment is processed correctly from both the manufacturer's and user's perspectives and that the necessary paperwork has been verified [10].

Documentation for operational validation includes:

- finalized and approved operations (functions testing)
- certified calibrations
- system stability test results
- S.O.P.s applications

Performance Qualification

Performance qualification ensures that the equipment consistently complies with the given specifications, which are suitable for daily/routine use. It is a formalized verification process that guarantees all elements of the building, its functionality, and the equipment's performance meets predetermined criteria from the user requirement specification (URS) and the manufacturer's requirements. Performance qualification is carried out under controlled settings, much like daily sample analysis, when equipment is used or functions are performed. When equipment is utilized or tasks are performed, it is done every day (or at least once a week). System suitability testing is another name for it, and it is carried out more regularly than operational qualification. The stability of each component of the entire system, in addition to the equipment's performance, determines the frequency of testing, which improves the outcome of the analysis [11].

Documentation for performance validation includes:

- performance qualification report
- process stability testing reports (long-term productivity)
- acceptance of the product record (costumers reviews)
- actual product and process parameters documentations
- routinely performed test results documentation

Revalidation

Revalidation is done when the system and running hardware have undergone any kind of modification. Equipment revalidation is particularly beneficial in maintaining the validation condition of the individual pieces of equipment as well as the entire system. In accordance with government regulations, the revalidation method is additionally employed to routinely assess the validity of the validation [4, 5, 12].

Revalidation is further divided as follows:

- Periodic/scheduled revalidation.
- Revalidation after change/modifications.

Periodic revalidation is the term for the revalidation procedure required by the pharmaceutical industry that must be carried out at regular intervals, especially when the company modifies the formulas, practices, manufacturing systems, packaging, and support systems like the electricity/power supply, water supply, and steam. A different, highly qualified crew will attend in the case of equipment revalidation, with the analyst coming from the manufacturer's side. Validation is necessary even after small product changes since even small changes can have a big influence on the product's quality. Sometimes, even during the initial validation, operational and performance tests were repeated.

The following is the alterations for revalidation are necessary these are as follow:

- modification in raw material
- modification in manufacturing process
- modification in equipment/system
- alteration in supporting systems
- alteration in packaging materials

Validation Maintenance Phase

The procedure known as Maintenance Qualification (MQ) looks at and confirms the acceptability of maintenance controls in order to guarantee the equipment's integrity systems. A recorded periodic evaluation is required for maintaining processes, systems, and equipment. It is a standard technique that ensures that contamination or structural problems in the equipment won't damage the product's identification, safety, or other attributes. The maintenance qualification process includes regular maintenance and necessary repairs [13].

Detailed maintenance contracts and a list of authorized services engineers are included in the documentation for maintenance qualification.

Applications of Equipment Validation

The following are the importance of equipment validation in pharmaceutical industries: [14].

- By minimizing rejects, reworks, and downtime, as well as by reducing the risk of regulatory non-compliance, equipment validation reduces costs.
- Customer happiness is really high.
- Testing and calibrations for analytic systems are done.
- It also lessens testing of the finished product and in-process goods.
- Additionally, raise staff awareness.
- Make equipment upkeep simple.
- Ensure a quicker and more reliable start-up for new equipment.
- Assist in creating the validation master plan for the facility.
- The validation documentation may be used as a presentation during an inspection (as evidence in a court of law).

References

1. Sharma PP (2016) Validation in pharmaceutical industry, 6th edn. Informa Healthcare Publication, UK, pp 51–70
2. Willing SH (1997) Good manufacturing practices for pharmaceuticals, 5th edn. Marcel Dekker Publication, New York, pp 66–80
3. Haider SI (2001) The pharmaceutical master validation plan: the ultimate guide to FDA, GMP, and GLP compliance hardcover-illustrated. CRC Press, Cleveland, pp 115–125
4. Nandhakumar L, Dharmamoorthy G, Rameshkumar S, Chandrasekaran S (2011) An overview of pharmaceutical validation. Quality assurance view point. Int J Res Pharm Chem 1:1003–1014
5. Todde S, Kolenc Peitl P, Elsinga P, Koziorowski J, Ferrari V, Ocak EM et al (2017) Guidance on validation and qualification of processes and operations involving radiopharmaceuticals. EJNMMI Radiopharm Chem 2:8

6. World Health Organization (1992) Equipment and its qualification, Rutendo Kuwana technical officer. World Health Organization, Geneva, pp 14–96
7. U.S Food and Drug Administration (2009) Guidance for industry, process validation: general principles and practices. U.S Food and Drug Administration, Washington, DC
8. The United States Pharmacopoeia Convention (1995) Validation of compendia methods. USP–NF, USP Convention Inc., Rockville, MD, pp 1612–1710
9. Chapman KG (1991) A history of validation in the United States. Part I Pharm Technol 15:82–96
10. Health Products and Food Branch Inspectorate (2009) Validation guidelines for pharmaceutical dosage form. Health Products and Food Branch Inspectorate, Canada, pp 8–13
11. Von Doehren PJ, St. John FF. (1982) An approach to the characterization and technology transfer of solid dosage form processes. Pharm Technol 6:139–156
12. Mantry S, Ramya I, Anil Kumar S (2014) An overview: pharmaceutical validation. Int J Innovative Pharm Sci Res 2:2476–2497
13. Mahar P, Verma A (2014) Pharmaceutical process validation: an overview. Int J Appl Pharm Biol Res 3:243–262
14. Rivera RA (2009) Validation theory and its application. Telstar Manufacturing Corporation, Laguna, pp 12–16

Chapter 12
Cleaning Validation for the Pharmaceutical, Biopharmaceutical, and Nutraceuticals

Abstract By avoiding cross-contamination, cleaning and cleaning validation have the greatest potential to lower patient risk. Ineffective cleaning from previous product batches, cleaning chemicals, or other extraneous material integrated into the product created by the procedure might result in adulterated product. Cleaning validation is important because we work with complicated biotechnology products and powerful, intricate pharmaceutical substances. Every component of cleaning validation will be covered in this section.

Keywords Validation · Cleaning validation · Pharmaceuticals

Introduction

Cleaning Validation

There is proof to support the claim that one can reliably clean a system or piece of equipment to acceptable and stated limits [1, 2]. The main goal of validating a cleaning technique is to confirm that it conforms to applicable legislation, both federal and state. The discovery and rectification of previously identified potential defects that could endanger the safety, effectiveness, or quality of subsequent batches of drug product produced by the apparatus is the main benefit of carrying out such validation work [2].

Objective

Cleaning procedure efficiency in terms of eliminating product residues, degradation products, preservatives, excipients, or cleaning agents, as well as microbial contamination, is checked during cleaning validation. For the following reasons, cleaning procedures need to be confirmed:

S. S. Shukla et al., *Pharmaceutical Calibration, Validation and Qualification: A Comprehensive Approach*, https://doi.org/10.1007/978-981-19-9002-1_12

- Pharmaceutical items and API can be contaminated by other pharmaceutical products, cleaning products, and microbiological contamination.
- The issue is the same: Assurance that the equipment is clean and that the product quality and safety are maintained. It is a regulatory requirement in the production of pharmaceutical items.
- From the perspective of internal control and compliance, it also ensures production quality.
- To retain the product's quality.
- The ability to reuse the apparatus [1].

Need for Cleaning Validation

To ensure that cleaning methods are efficient and that there are no risks associated with the pollution of detergents' active ingredients.

Why Cleaning Validation [1]

- Initial equipment/process qualification.
- A substantial change to a cleaning method. a substantial change to a composition.
- The formulation has undergone a considerable alteration.
- A modification to the cleaning process.
- Change the cleaning substance.

In Case of Drug Products

Different cleaning techniques pharmaceutical manufacturing processes may involve scenarios like batch-to-batch changeover cleaning. From one product to the next, clean. Depending on the manufacturing stage and the kind of the next manufacturing step to be performed in the same equipment, various cleaning processes may be used in non-dedicated drug product production facilities.

There are consequently two levels of cleansing, as detailed below [1].

Level 1 Cleaning

This is done to identify between various production runs of the same product.

Level 2 Cleaning

Even if the same product is planned for the following campaign, this is used in between batches of separate items being created and/or towards the conclusion of a production campaign.

Cleaning Validation Master Plan

Master plan should [1]:

- Describe the site, building, or region within the Master Plan's jurisdiction.
- Describe the typical production procedures that will be carried out in the region, as well as the dosage forms that will be produced.
- Describe the cleaning techniques that will be employed (e.g. automated Clean-In-Place or Clean-Out-of-Place, semi-automated cleaning, or manual cleaning).
- The responsibilities of the several departments for cleaning validation activities.
- Assign the cleaning validation programme the fewest requirements possible [1–3].

Elements of Cleaning Validation

- residue selection
- equipment characterization
- cleaning agent selection
- limits calculation
- product grouping
- equipment grouping
- cleaning procedure
- sampling
- analytical methods
- validation protocol
- validation report

Residue Identification

There are a number of residues to consider when doing cleaning validation:

- API
- constituents of the cleaning agent

- preservatives
- precursors or starting materials
- intermediates
- processing aids
- media
- buffer
- cellular debris or metabolites
- particulate
- bioburden
- endotoxin
- viral particles
- TSE
- excipients
- colourants, dyes, flavours, or fragrances

Equipment Characterization

In addition to removing residues, cleaning validation comprises verifying that each and every piece of process equipment has been cleaned in accordance with appropriate standards.

It is frequently known as a train-based method.

A group of machines known as an equipment train is used to transport a product or a group of products through the production process.

The equipment should be defined in such a way that all of its design elements are clear in order to establish whether or not it is cleanable.

Equipment characterization could help cleaning validation programmes in a number of ways:

- Recognize cleaning challenges and make sure they are addressed in the cleaning techniques employed to encourage more successful cleaning procedures.
- Making use of tools to locate high-risk and challenging-to-clean locations in order to choose sample sites.
- Determine which building materials will be examined in sampling recovery investigations and which ones won't.
- Separate materials that will be used for a specific product or that will be thrown away after a manufacturing process.
- Check that all construction components can withstand the temperature and cleaning solutions being used throughout the cleaning process [1–3].

Cleaning Agent Selection

All cleaning processes depend on the principle of TACT and WINS.

TACT

Time, Action, Concentration/Chemistry, and Temperature, or TACT, are variables that must be controlled in any cleaning procedure, whether it is manual, somewhat automated, or fully automated. When one TACT parameter is changed, the other parameters will follow suit. However, in all cases, extensive understanding of WINS is necessary for correct TACT parameter balancing [1–3].

WINS

Person, Surface, Nature, and Water. The acronym WINS stands for the factors that affect the removal of soil from the surface, and each factor might have an impact on your ability to employ TACT in a certain circumstance. Chemicals used for cleaning are broken down into a number of categories, including water, solvents, common chemicals, and specially developed cleaning agents.

Grouping of Equipment

All machinery must be equal in terms of position or function within the manufacturing process, utilized to produce goods from the same product category, cleaned using the same cleaning agent, and cleaned using the same cleaning technique [1–3].

Product Grouping and Equipment Grouping

It is a technique for figuring out whether two pieces of equipment or merchandise are comparable or equivalent for the purposes of cleaning validation. A worst-case area of the instrument or site is chosen for cleaning validation when comparing similar instruments or sites. When identifying equivalent, any section of the instrument or site may be selected as a representative of any other area of the instrument or location [1–3].

The term "bracketing", which is used in the EU GMP Annex's section on cleaning validation, means the same as "grouping"; however, it may come with a heavier load for testing population extremes.

Grouping can be used to either reduce some of the innumerable possible product and equipment combinations for study or to simply prioritize cleaning validation investigations.

Grouping for Products

The same piece of machinery must be used for all production, and the same cleaning agent must be used on all items. The same process of cleaning was employed. The parameters utilized to classify goods include similar patient risk levels (e.g. therapeutic indication, potency, and toxicity for drugs, devices, nutraceuticals, and cosmetics) [1–3]. To satisfy the lowest limit of the entire product group and geometry, materials of construction, and capacity, cleaning validation of similar formulas, manufacturing techniques, and geometry, materials of construction, and capacity must constantly be carried out [4].

Cleaning Procedures

Develop a standard cleaning procedure for every piece of machinery and procedure. It is essential to carefully review the equipment design in order to eliminate product residues. When using cleaning techniques, bear the following considerations in mind:

A. Equipment parameters to be evaluated.
   - Identifying the equipment to be cleaned.
   - "Difficult to clean" parts.
   - Material properties.
   - Ease of disassembly.
   - Mobility.

B. Residues to be cleaned.
   - Solubility of residues is limited by cleaning.
   - Campaign duration.

C. Cleaning agent parameters to be evaluated.
   - Preferable materials commonly used in the process.
   - Detergents available (detergents should be used sparingly unless absolutely necessary).
   - Solubility characters.

- Environmental factors.
- Consider for health and safety.

D. Cleaning techniques to be evaluated.

- Manual cleaning.
- CIP (Clean-in-place).
- COP (Clean-out-of-place).
- Semi-automated and automated procedures.
- Time concerns.
- Number of cleaning cycles.

Sampling Technique

Due to the design of the equipment being difficult to clean and the fact that these areas are inaccessible, sampling sites were chosen (i.e. their inaccessibility makes them difficult to clean). As a result, one must be cautious while choosing sampling locations. The two categories of equipment are hot spots and crucial locations. A hot spot is a place that is likely to get dirty and be challenging to clean during the manufacturing process. Critical areas are those that, if left dirty, will unavoidably lead to excessive contamination in the subsequent batch of exhibits. In cleaning validation, swab sampling, direct surface sampling, and rinse sampling are the most often utilized sample methods [5, 6].

Direct Surface Sampling

In order to evaluate whether the sample material interferes with the test, it is necessary to identify the type of sampling material used and its effect on test data. Therefore, it is crucial to check early on in the validation process that the sampling medium and solvent are suitable and simple to use [7–10].

This is done using FTIR or photoelectron emission methods. Using these methods, specific spectra obtained from residue left on the surface will be utilized to quantify the surface quality directly [11, 12].

Swab Sampling

The analyte must typically be removed physically by cleaning the surface and using materials that absorb substances. Swabs should be compatible with the active ingredients and not interfere with the assay. They shouldn't let the chemical deteriorate in any manner. The chemical should be well-soluble in swabbing solutions, which shouldn't encourage deterioration [12].

Rinse Sampling

Rinse sampling collects the sample as a final rinse or as a rinse used specifically for collecting a validation sample and has no mechanical effect on the surface. Obtaining and testing rinse samples for residual active component is a common technique for assessing cleanness. Although it calls for careful management of the washing solvent, contact duration, and mixing, this technique is generally quite convenient. The solvent should be selected based on the solubility of the active ingredient and should either replicate a subsequent batch of the product or at the very least provide enough solubility [13].

Placebo Sampling

Both a potential cleaning method and a potential sample strategy have been identified as the placebo. The placebo material comprises all of the customary excipients but lacks the active component. In order to clean the clean system, the placebo batches were also run through the same line. According to the placebo theory, as it travels along the same pathway as the substance, it may remove any remaining substance along those pathways. It is widely employed to examine a system's cleanliness [13].

The main factors are: The excipients' solubility in the placebo; the placebo's proper contact time for getting a representative sample. Covering the placebo in-process channels makes sure that every piece of equipment has the placebo taken out of it. The ratio of residue to placebo must fall within a detectable range, and the uniform distribution of residue throughout the placebo guarantees sample detection throughout the whole placebo. The swabbing method is the preferred sampling method and the one that regulatory bodies regard to be the most acceptable.

Analytical Techniques

Numerous variables affect which analytical technique is best to use. Choosing the requirements or parameters that will be measured is the most important step. The limit should always be decided upon before using an analytical tool [5, 14]. There are two ways to go about things: a method that is particular and a method that is not. Using a specialized method, one can identify unusual compounds when there are possible contaminants present. A nice illustration is HPLC. Processes that identify any substance that induces a reaction are referred to as non-specific procedures. Example variables include conductivity, pH, and total organic carbon (TOC) [15].

Additional Techniques

In addition to the methods mentioned above, the biopharmaceutical industry uses a wide range of strategies. In order to assess the quality of surfactants, TLC is frequently utilized. Atomic absorption spectroscopy is used to identify inorganic contaminants. Bioluminescence is advantageous to biologicals. ATP-bioluminescence is frequently employed in this kind of research. Also covered are limulus and the ELISA (Enzyme-Linked Immuno Sorbent Assay) [16, 17].

Validation Protocols

A Validation Protocol is necessary to outline the particular items and actions that will make up a cleaning validation study. The overall Cleaning Validation approach for the product line, equipment type, and complete site should be outlined in a company's Master Validation plan. The protocol must be created before to the study's start and must either refer to or include the documents required to give the following information:

Background

- purpose of the validation study
- scope of the validation study
- responsibilities for performing the validation study
- sampling procedure to be used
- testing method to be used
- acceptance criteria
- change control
- approval of protocol before the study
- deviations [16]

Validation Reports

A validation report is required to present the study's findings and conclusions and to gain approval. The following items should be included in the report:

- A description of or a citation for the cleaning, sampling, and testing procedures a summary or citation of the physical and analytical test results, together with any significant findings.
- Any recommendations based on the findings or pertinent data gathered throughout the study, including, if necessary, revalidation practices. Conclusions on the validity of the procedure(s) being validated and the acceptability of the results are approved, and any deviations from the protocol are reviewed.
- When producing additional batches of the product won't likely occur for a while, it is advised to prepare interim reports batch by batch until the cleaning validation research is complete.
- Following validation, the report should conclude with an adequate level of verification.

A successful cleaning validation maintenance technique [17, 18]. The cleaning techniques will have proven to sufficiently and consistently remove chemical and detergent residues from equipment surfaces throughout the study to meet the pre-established requirements if at least three cleaning validation runs are conducted and the findings satisfy the acceptance criteria. However, time and other factors can affect the cleaning programme's effectiveness and frequency [19–21]. They are

- operator inconsistency
- equipment age and repair
- monitoring programmes that could produce findings that aren't representative, changes to the product, equipment, and process
- operator variability [21]

Conclusion

Cleaning validation, in conclusion, is the process of obtaining and documenting adequate evidence to demonstrate the success of a cleaning method. Cleaning is closely related to the pharmaceutical product's safety and purity; thus, it becomes the most critical and primary task. As a result, the regulatory requirement necessitates the implementation of an effective cleaning programme. This chapter goes through everything you need to know about cleaning validation, including residue selection, validation acceptance criteria, different levels of cleaning, cleaning technique, sample procedure, product grouping and equipment characterization, and cleaning chemical selection.

References

1. Rebecca Brewer Dober Group (2007) Cleaning validation for the pharmaceutical, biopharmaceutical, cosmetic, nutraceutical, medical device, and diagnostic industries validation of

pharmaceutical processes validation of pharmaceutical processes. 3rd ed. Midlothian, IL. pp 491–518
2. Sharma A, Vaghela J.S, Sharma P, Sharma B, Agrawal M, Jagiwala S. Cleaning validation in pharmaceutical industry: a comprehensive approach; pharmatutor-art-1755
3. Madsen RE (1998) Points to consider for cleaning validation. Technical report no. 29. PDA J Pharm Sci Technol 52(6):–1, 53
4. Fredrick R (2004) The basic facts of cleaning validation. Pharmainfo.net. http://www.pharmainfo.net/reviews/basic-facts-cleaning-validation
5. Fourman GL, Mullen MV (1993) Determining cleaning validation acceptance limits for pharmaceutical manufacturing operations. Pharm Tech 17(4):54–60
6. Vitale KM (1995) Cleaning validation acceptance criteria. In: 15th Annual Pharm Tech conf, East Brunswi, NJ, vol. 95. pp. 18–21
7. Brewer R (1996) Regulatory aspects of cleaning validation. ISPE seminar, Rockville, MD, pp 6–8
8. Narayana Murthy D, Chitra K, Fourman GL, Mullen MV (1993) Determining cleaning validation acceptance limits for pharmaceutical manufacturing operations. Pharm Tech 17(4): 54–60
9. Rao BSS (2012) Cleaning validation of albendazole tablet 400 mg. Pharma Innovations 1(4): 76–94
10. Kumar S (2012) A review on concept of cleaning validation in pharmaceutical industry. Int J Pharm Sci Res 3(7):17–19
11. Kumar VS, Sanjeev T (2012) Overview of cleaning validation in pharmaceutical manufacturing unit. Int J Adv Res Biol Sci 1(2):154–164
12. Narayana Murthy D, Chitra K (2013) A review article on cleaning validation. Int J Pharm Sci Res 4(9):3317–3327
13. Kaiser HJ, Tirey JF, LeBlanc DA (1999) Measurement of organic and inorganic residues recovered from surfaces. J Valn Tech 6(1):424–436
14. LeBlanc DA (1998) Establishing scientifically justified acceptance criteria for cleaning validation of finished drug products. Pharm Tech 22(10):136–148
15. Inampudi P, Lombardo S, Ruezinsky G (1996) An integrated approach for validating cleaning procedures in biopharmaceutical manufacturing facilities. Ann N Y Acad Sci 782:363–374
16. Rowell FJ, Miao ZF, Neeve RN (1998) Pharmaceutical analysis nearer the sampling point, use of simple, rapid on-site immunoassays for cleaning validation, health and safety, and environmental release applications. J Pharm Pharmacol 50:47
17. Hwang RC (2000) How to establish an effective maintenance program for cleaning validation. Pharm Tech 24(1):62–72
18. McCormick PY, Cullen LF (1993) Cleaning validation. In: Berry IR, Nash RA (eds) Pharmaceutical process validation, 2nd edn. Marcel Dekker, New York, pp 319–349
19. Lakshmana P (2010) Cleaning validation and its importance in pharmaceutical industry. Pharm Times 42(7):21–25
20. James A (1992) Points to consider in the validation of equipment cleaning procedures. J Parental Sci Tech 46(5):163–168
21. Phillips GF (1989) The need for impurity control. Pharm Ind 51(11):1282–1286

Part VI
Validation of Dosage Forms

Chapter 13
Validation of Pharmaceutical Dosage Forms

Abstract Quality has always been a crucial prerequisite when taking any goods into consideration. As a result, the highest quality standards must be used in the manufacturing of pharmaceuticals. A well-organized, skilled, and proficient production operation is essential to building quality into the goods. Validation is a component of the quality assurance programme. Process Validation is the term used to describe the validation of various process phases. In order to build a scientific basis for the process' ability to consistently produce high-quality medicinal material, it thus requires the collection and evaluation of data from the process' inception stage through all stages of production. The goal of validation is to confirm that quality is developed throughout the entire process and is not checked in the last stages. According to CGMP, Process Validation is a crucial and essential component of Quality Assurance (QA). Together, validation and quality assurance work to guarantee the products' methodical quality. Validation studies must be carried out in accordance with established protocols and are required under GMP. Different validation processes apply to various dosage types. This chapter's goal is to provide a comprehensive description of the introduction and information concerning Process Validation of the pharmaceutical manufacturing process. It will thus provide in detail the validation of each and every step in the manufacturing process through the process.

Keywords Pharmaceutical dosage forms · Tablets · Capsules · Liquid · Regulatory basis · Protocol · Quality · Validation

Introduction

Directives for Process Validation of Solid Dosage Forms

Numerous factors need to be taken into consideration while the development and validation of solid dosage forms. A broad and large indication of these validation criteria is being depicted in the following table which acts as checklist/guideline.

It is provided for tablets and dry-filled capsules for addition in an in-depth validation programme. Some of the unit operations may not be relevant and pertinent

S. S. Shukla et al., *Pharmaceutical Calibration, Validation and Qualification: A Comprehensive Approach*, https://doi.org/10.1007/978-981-19-9002-1_13

Table 13.1 Check list of validation and documentation

Range of cGMP	Validation and documentation
Foreword	For establishing of QA and PV functions
Organization and personnel	For establishing the facility of installation and qualification
Buildings and facilities	For plant and facility installation qualification, maintenance and sanitation microbial and pest control
Equipment	For installation and qualification cleaning methods
Air and water quality	For water treatment and steam systems air heat and vacuum handling
Control of raw material, in-process material product	For incoming components and manufacturing non-sterile products
Production and process controls	For process control systems of instruments and computers
Packing and labelling controls	For depyrogenation, sterile packing, filling, and closing
Holding and distribution	For facilities
Laboratory controls	For analytical methods
Records and reports	For computer systems
Returned and salvage drug products	For batch processing

Table 13.2 Title page protocol in the industry

Name of the company	
Process validation protocol	
Product:	Page no: 1 of
Protocol no.:	Version no:
Product name:	
Label claim:	
Master formula record (MFR) no.:	
Batch manufacturing record (BMR) no.:	
Effective date:	

for every solid dosage form (e.g. direct compression tablets and uncoated tablets) [1–5](Table 13.1).

Protocol for Process Validation

The protocol for Process Validation has been depicted in detail in the following tables (Tables 13.2, 13.3, 13.4, and 13.5) which are as follows:

Table 13.3 Protocol approval

	Prepared by	Checked by			Approved by
Signature					
Date					
Name					
Department	Quality Assurance (QA)/ Research and development (R&D)	Officer R&D	Officer Production	Officer Quality Control	Head QA

Table 13.4 Table of contents for protocol

Table 13.5 Validation team and responsibilities

Department	Designation	Responsibility
Research and development (R&D)	Executive/officer	• Coordination of the complete validation process by scheduling of meetings, discussions with production, quality control, and quality assurance • Preparation of preliminary validation protocol, master formula record (MFR), monitoring of the process, compilation and analysing the data and test results and preparation of the final report • Review of the preliminary validation documents
Quality assurance	Officer	• Coordination of the entire validation process by scheduling meetings, discussions with the team • Preparation of the validation protocol, monitoring of the process, compilation and analysis of data and test results, and preparation of the final report • Review of validation documents
Production	Officer	• Participation in performance of the validation steps during the manufacturing processes • Assistance in the data collection
Quality control	Officer	• Testing and reporting of the test results
Quality assurance	General manager	• Approval of the process validation protocol and report • Review of validation documents • Approval of the process

Steps for Validation and Acceptance Criteria

The following are the steps as mentioned in Table 13.6 which are employed in the industry for validation of tablets through wet granulation process.

Industrial Process Assessment and Selection for Tablets

Determination of the Unit Operations Required to Fabricate the Tablets Is as Follows

Mixing or Blending

Materials having analogous physical properties will be unproblematic to form a homogeneous mix or blend and will not separate out as readily as materials with large differences [3, 14, 16, 18] (Fig. 13.1).

Parameters to be considered are as follows:

(a) *Mixing or blending technique*

The different techniques like diffusion (tumble), convection (planetary or high intensity), or pneumatic (fluid bed) can be employed either to mix or blend the materials. The technique needs to be determined which is obligatory for the

Table 13.6 Steps for validation and the acceptance criteria

S. No.	Steps	Variables	Parameters	Acceptance criteria
1	Dry mixing	Time Impeller speed	Mixing time and speed	• Mixing time min. • Impeller speed: (slow/medium/high) ± 5RPM • Content uniformity: 90%–110% • RSD: ± 5%
2	Binder preparation and addition	Time Temperature, solvent used	Mode and time of addition	• Depending upon the formulation
3	Kneading	Time Impeller speed and chopper speed	Mixing time and speed	• Impeller speed: (slow/medium/high) • Chopper speed: (slow/medium/high) Depending upon the formulation
4	Drying	Inlet/outlet temperature and time	Inlet/outlet temperature and drying time	• Initial drying °C. • Drying time min. • Final drying °C ± 5 °C • Loss on drying %. Below 3% or depending on formulation
5	Lubrication	Time Blender/granulator speed	Mixing time and speed	• Mixing time in min • Speed: Slow rpm • Content uniformity: • Physical parameters: For information
6	Compression	Pressure and turret speed	Machine speed and compression pressure	• Average weight: mg ± 5%, 7.5%, 10% • Uniformity of weight mg: Thickness in mm • KN or kg/cm^2 • Disintegration time: NMT in min • Friability: NMT % w/w • Assay: as per the label claim • Dissolution: %
7	Coating	Pan speed and spray rate	Pan speed inlet and outlet temperature spray rate	• Average weight mg ±5% • Weight of 20 tablets mg • Thickness: mm • Disintegration time: NMT in min

(continued)

Table 13.6 (continued)

S. No.	Steps	Variables	Parameters	Acceptance criteria
				• Assay: As per the label claim • Dissolution: %

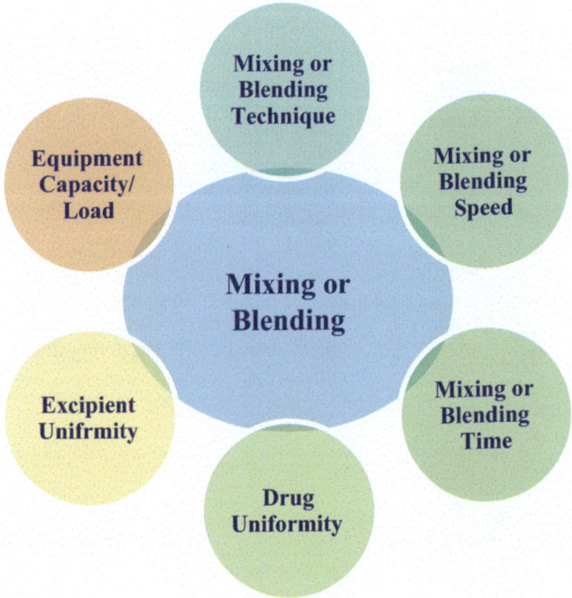

Fig. 13.1 Parameters for mixing or blending

formulation or process objective. It may be different, depending on whether the drug and excipients are mixed by a direct compression formulation or by addition of the lubricant (e.g. magnesium stearate) to the granulation.

(b) *Mixing or blending speed*

The intensity (low/high shear) and/or speed (low/high/optimal shear) (rpm) of the mixing or blending is determined. Mixing of the drug and excipient requires further intense mixing than adding up the lubricant to the final blend.

(c) *Mixing or blending time*

The mixing or blending required to acquire a uniform mixture is determined. The mixing or blending time will be dependent variable on the mixing or blending technique employed and speed ascertained. Experiments ought to be done in order to determine that the overmixing of the materials which ultimately results in demixing or separation of the materials. Demixing can take place due to some physical property differences, for example, particle size distribution and density. For example, demixing can come about in a direct compression formulation in which the drug substance is micronized (5 microns) and the excipients are granular (500–1000 microns).

(d) *Drug uniformity*

Content uniformity is generally executed in order to determine the drug uniformity throughout the mix or blend. Representative samples should be occupied during the mix or blend. The sampling modus operandi and handling of the materials are vital in obtaining the convincing content uniformity results. Separation of the sample can occur by over-handling, consequential in imprecise and erroneous results. For the absolute blend (blend aforementioned to compression), the sample taken ought to be comparable to the weight of a single tablet.

(e) *Excipient uniformity*

Besides drug uniformity, excipients necessitate to be standardized in the granulation or blend. Two key excipients are:

i. *Lubricant*

The lubricant desires to be disseminated consistently in the mixture/granulation for the high-speed solidity operation. Irregular division of the lubricant can upshot in picking and sticky problems during compression. It can also escort to tablet performance problems such as low-down dissolution due to disproportionate lubricant in some tablets.

ii. *Colour*

The colourant(s) need(s) to be consistently disseminated in the mixture so that the tablets have a homogeneous manifestation (e.g. colour, hue, and intensity). The colouring agent ought to be pre-screened or more regularly dispersed in the blend preceding to compression in order to avoid speckling or shading of the colour.

(f) *Equipment capacity/load*

The bulk density of materials or granules will influence the capacity of the equipment. If an excipient in the formulation concerns the density of the final blend to a superior coverage than any other ingredient, then a fine-guarded density requirement for that excipient may be warranted. Test diverse-sized loads in the mixer/blender, for example, 30, 50, and 70% of working volume for best possible mixing or blending. Under-charging or over-charging a blender can upshoot in poor drug or tablet lubricant circulation.

Wet Granulation

The following criteria needs to be looked upon such as type of wet granulation technique to be employed along with the shear either low shear, for example, Hobart, or high shear, for example, Diosna, GEI-Collette or fluidized bed, for example, Glatt, Fluid Air [6].

Each technique will fabricate granules with diverse physical properties and will necessitate for extra monitoring of altered processing parameters.

The wet granulation parameters to be taken into consideration during development and validation are as follows:

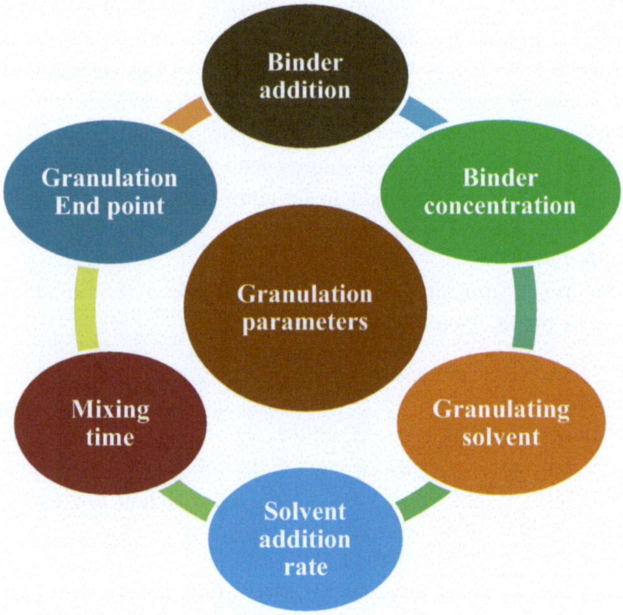

Fig. 13.2 Granulation parameters

(a) *Binder addition*

It should be predetermined about the binder addition to be added as granulating agent or like other excipients added to the formulation. Addition of the dry binder dry evades the requirement in order to resolve the finest or optimum binder concentration and a separate fabrication for the binder solution (Fig. 13.2).

(b) *Binder concentration*

The optimum binder concentration needs to be predetermined for the formulation. If the binder is to be sprayed, the binder solution ought to be sufficiently diluted in order to get it pump easily via spray nozzle. It should also be adequately and amply concentrated to outline granules without over wetting the materials.

(c) *Amount of binder solution/granulating solvent*

It should be pre-estimated about the binder or solvent solution required in order to granulate the material. If too much binder or solvent solution is added, it will lead to over wetting of the materials and which will in turn prolong the drying time. The quantity of binder solution is interrelated to the binder concentration.

(d) *Binder solution/granulating solvent addition rate*

The rate or rate range needs to be determined at which the binder solution or granulating solvent will be added to the materials. It should also be taken into consideration that the granulating solution either needs to be dumped into the mixer or it has to be metered in at an explicit rate.

(e) *Mixing time*

The mixing time of the material needs to be predetermined, in order to ensure the proper formation of granules. Steps need to be pre-estimated for the stopping of mixing, addition of binder or solvent solution or any other additional mixing (if required). Granulation that is mixed for longer time can form incomplete or weak granules. These granules may have reduced flow and compression properties. On the contrary, the over mixing in the granulation can direct to the harder granules and a subordinate dissolution rate.

(f) *Granulation end point*

The end point of the granulation needs to be determined. It is either determined or controlled by granulation endpoint equipment such as ammeter or wattmeter. It is also controlled by specifying some critical processing parameters. Such as a drug or excipient mixture may be granulated either by adding a predetermined amount of water which acts as granulating solution at a definite rate. The granulation is concluded after mixing for a set time after the water has been added [6].

Wet Milling

The requirement of wet granulation to be milled is done in order to break up the lumps which will enhance the drying of the granulation. Wet granules having a spacious aggregate range can lead to inefficient drying (long drying times and partially dried large granules or lumps) [7] (Fig. 13.3).

Factors to be considered are:

(a) *Equipment size and capacity*

The mill should be outsized adequate to de-lump the complete batch within a practical time period in order to curtail the manufacturing time and avert the material from drying all through this operation.

(b) *Screen size*

The screen requirement is small and enough to de-lump the material, but not very small so that it may cause disproportionate heating of the mill, which will ultimately result in drying of the granulation.

(c) *Mill speed*

The speed should be adequate to resourcefully de-lump the material with no straining of the equipment.

(d) *Feed rate*

The feed rate of the wet granulation is interconnected to the screen size, mill size, and speed [7].

Fig. 13.3 Wet milling factors

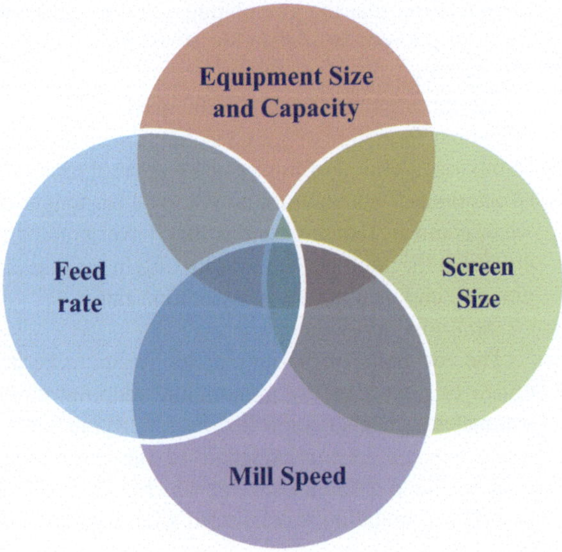

Drying

The type of drying technique, e.g. tray, fluid bed, and microwave, essential and vital for the formulation is required to be dogged and justified. The type of method employed may depend on certain factors such as drug or formulation properties and equipment availability. Altering dryer techniques could influence such tablet properties as hardness, disintegration, dissolution, and stability [8–10]. The optimal moisture content of the dried granulation ought to be determined. High moisture content can result into the following:

(a) tablet picking or sticking to tablet punch surfaces
(b) reduced chemical stability as a consequence of hydrolysis

An over-dried granulation could upshot in reduced hardness and friability. Moisture content analysis can be performed by employing the predictable loss-on-drying techniques or such state-of-the-art techniques as near-infrared (NIR) spectroscopy (Fig. 13.4).

The parameters to be taken on to consideration are as follows:

(a) *Inlet/outlet temperature*

The inlet temperature refers to the temperature of the inward bound air to the dryer, while the outlet temperature refers to the temperature which leaves the unit. The inlet temperature is vital and significant to the drying efficiency of the granulation and should be set elevated enough to capitalize on drying without affecting the chemical as well as physical stability of the granulation. The outlet temperature is a pointer or marker of the granulation temperature and will

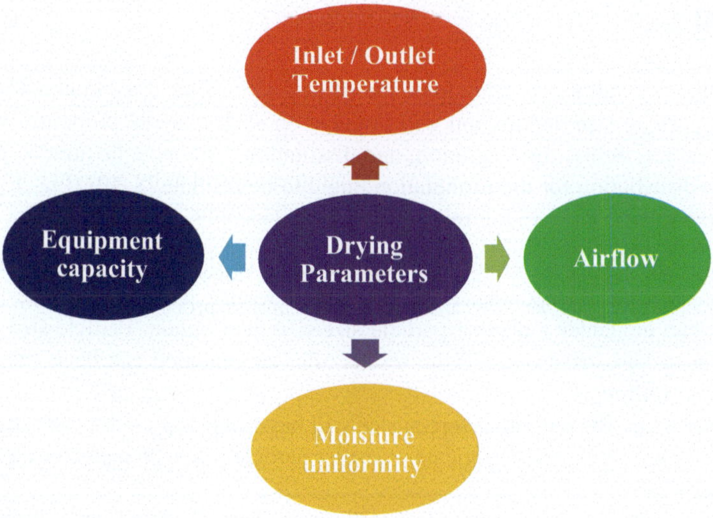

Fig. 13.4 Parameters for drying

augment towards the inlet temperature as the moisture content of the granulation reduces.

(b) *Airflow*

There ought to be adequate airflow in order to make certain the removal of moisture loaded air from the wet granulation. Inadequate airflow could protract drying and thus have an effect on the chemical stability of the drug. Thus, airflow and the inlet/outlet temperature are interconnected factors and should be taken into consideration together.

(c) *Moisture uniformity*

The moisture content could be different within the granulation. Heat uniformity of the dryer (e.g. tray), amount of granulation per tray, and partial fluidization of the bed are factors that could have an effect on the moisture uniformity of the granulation.

(d) *Equipment capability/capacity*

The load that can be capably dried within the unit is ought to be known. An outsized load will be required for more moisture content to be removed on drying and thus will have an affect the drying time. In the case of fluidized bed drying, a maximum dryer load is referred to as that load above which the dryer will not fluidize the material.

Milling

The milling operation will trim down the particle size of the dried granulation. The ensuing particle size distribution will impinge on such material properties such as flow, compressibility, disintegration, and dissolution. The most favourable particle size/size distribution for the formulation ought to be resolute [8–10] (Fig. 13.5).

Factors to be considered in milling are as follows:

(a) *Mill type*: The type of milling is imperative to determine along with other variants such as dependence on methods to reduce particle size. The type of mill can engender a diverse particle size/size distribution. Particle size testing ought to be carried out and the results after examination can have substituted type of milling.
(b) *Screen size*: The particular screen size will have an effect on the particle size. A smaller screen size will fabricate a smaller particle size with a greater number of fines.
(c) *Mill speed*: The optimal speed of milling needs to be determined, higher mill speed will correspond to smaller particle size and wide particle size distribution. It can also engender more heat to the product; this depends on the screen size and feed rate, which could impinge on the stability of the product.
(d) *Feed rate*: The feed rate is dependent on the mill capacity, screen size, and mill speed.

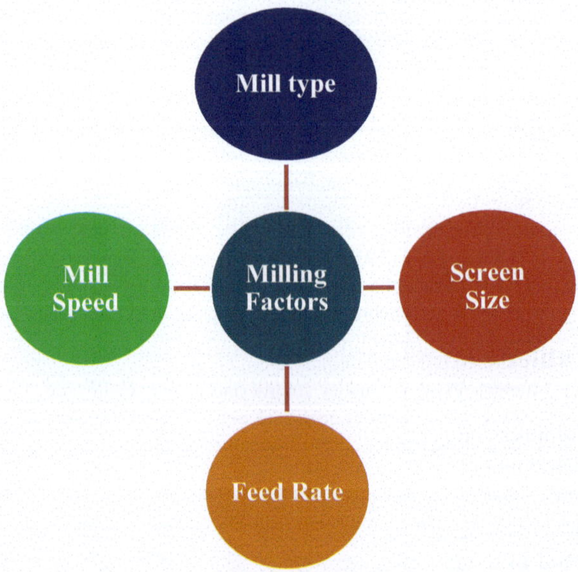

Fig. 13.5 Factors for milling

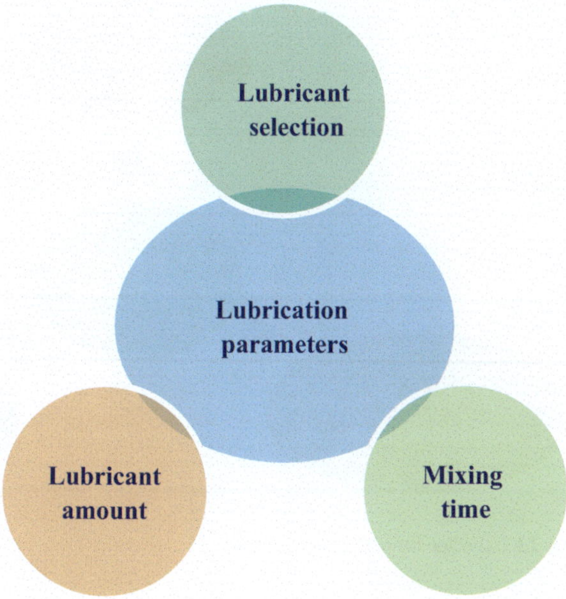

Fig. 13.6 Lubrication parameters

Lubrication (Fig. 13.6)

(a) *Selection of lubricant:* Lubricant is selected, which will be used in the process with due consideration to its grade and compatibility.
(b) *Amount of lubricant added:* The amount of lubricant to be added is optimized which will be required as too much and too low quantity will pose some defects. Too much lubricant will outline hydrophobic layer on the tablet which will result in some dissolution problems.
(c) *Mixing time:* It is crucial as the material is mixed in order to ensure proper formation. It also takes into consideration about whether mixing should be stopped after the addition of the lubricant or some sort of additional mixing is required. Mixing not long enough form can result into problems like chipping, capping, etc. [11–13].

Tablet Compression

Compression is an essential measure in the production of a tablet dosage form. The materials being compacted are required to have a passable flow and solidity properties. The material readily flows from the hopper on to the feed frame and into the dies. Insufficient flow can upshot into rat holing in the hopper and separation of the blend in the hopper/feed frame. This ground for predicaments in tablet weight and

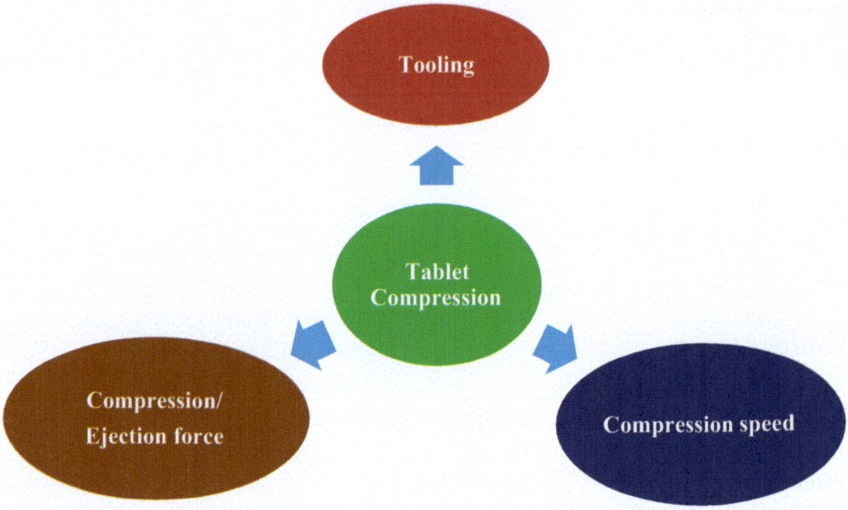

Fig. 13.7 Factor for tablet compression

content uniformity. As far as the compressibility properties of the formulation are concerned, it should be inspected on an instrumented tablet press [14, 15] (Fig. 13.7).

Factors to be taken into consideration during compression are as follows:

(a) *Tooling*

The shape, size, and concavity of the tooling ought to be scrutinized on the basis of the formulation properties and commercial specifications. For intagliated, embossed tablets, factors namely, the position of the intagliation on the tablet and the depth of intagliation and style should be scrutinized in order to ensure that picking of the intagliation during solidity or fill-in of the intagliation during coating does not crop up.

(b) *Compression speed*

The formulation should be compacted at a broad array of compression speeds in order to establish the operating range of the compressor. The competence of the material's flow into the dies will be resolute by scrutinizing the tablet weights. A force feeder is mandatory or required in order to ensure that enough material is fed into the dies.

(c) *Compression/ejection force*

The compression contour for the tablet formulation will necessitate to be unwavering in order to ascertain the most favourable firmness force so as to obtain the required tablet hardness. The particle size or size distribution or lubricant level may need to be attuned in order to have a vigorous process on a high-speed compressor.

The following are the in-process tests which need to be examined during the compression stage:

1. appearance
2. hardness
3. tablet weight
4. friability
5. disintegration
6. weight uniformity

Tablet Coating

Tablet coating can take place by diverse techniques such as sugar, film, or compression. Film coating has been generally widespread technique over current years. The coming section of the chapter will focus on the crucial part of tablet coating [14, 15] (Fig. 13.8).

Important parameters to be taken into consideration for tablet coating include the following:

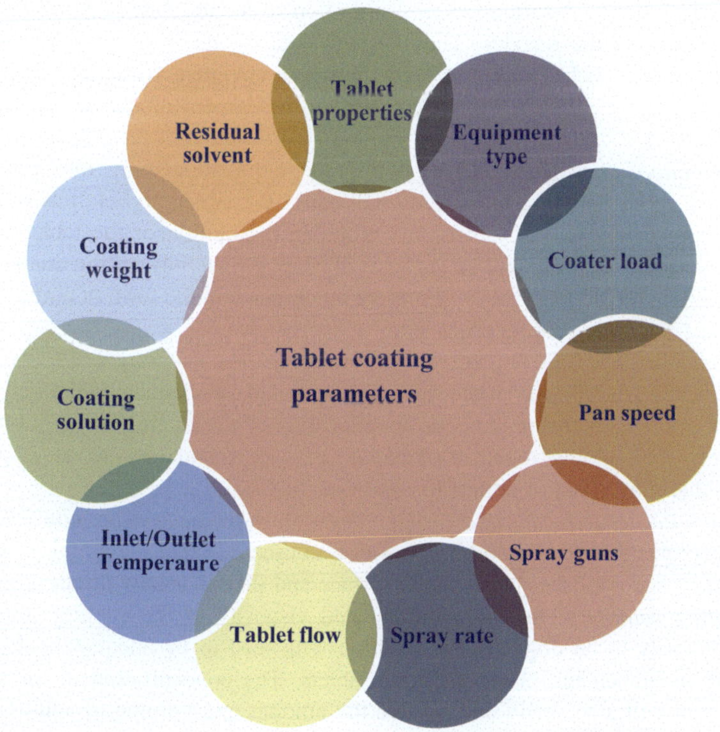

Fig. 13.8 Tablet coating parameters

(a) *Tablet properties*

Tablet properties like hardness, shape, and intagliation (whenever required) are basic to accomplish a good and decent film-covered tablet. The tablet is required to be sufficiently hard to hold up the coating cycle. In an event of tablet weakening, the tablets will have a nasty surface appearance.

For tablet shape, a round tablet will be effortlessly being coated than tablets will diverse sides or edges in light of the consistency of the surface. For intagliated tablets, the intagliation style and depth needs to be created in order to forestall fill-in or chipping of the intagliation.

(b) *Equipment type*

The type of coater will be required to be ascertained. Conventional or perforated pan and fluid bed coaters are probable and potential alternatives.

(c) *Coater load*

The acceptable tablet load range of the equipment needs to be ascertained. Too bulky pan load could be the reason for abrasion of the tablets. In the case of a fluidized bed coater, there may not be ample airflow to fluidize the tablets.

(d) *Pan speed:* The most favourable pan speed will be interconnected to other coating parameters, such as inlet temperature, spray rate, and flow rate.

(e) *Spray guns:* The number and category of guns should be dogged in order to have an efficient coat on to the tablets. The spray nozzles should be sized appropriately in order to ensure an even distribution over the tablet bed and to thwart congestion of the nozzles. The site and slant of the spray gun(s) should be positioned in such a way that it gets passable coverage. The guns when positioned too close to each other can lead to a segment of the tablets to be in excess wet.

(f) *Application/spray rate:* The best possible application/spray rate should be agreed upon. Spraying too speedy will cause the tablets to befall as over wet, resulting in thickets of tablets and probable dissolution of the tablet surface. Spraying too leisurely will be the basis of coating materials getting dried prior to sticking to the tablets. This will present an outcome in the form of a rough tablet surface and poor coating efficiency.

(g) *Tablet flow:* The flow or faction of the tablets in the coater ought to be examined in order to ensure appropriate flow. There should be adequate tablet bed movement in order to ensure the constant allocation of the coating solution on to the tablets. The adding together of baffles may be mandatory so as to provide passable movement of tablets for tablet coating.

(h) *Inlet/outlet temperature and airflow:* Both these parameters are consistent and interconnected and should be in position to ascertain that the atomized coating solution accomplishes at the tablet surface and then is hastily dried.

(i) *Coating solution:* The concentration and viscosity of the coating solution is obligatorily to be dogged. The solution will need to be adequately diluted in order to drench the material on the tablets. The concentration of the coating solution will also verify and resolve the amount and volume of solution to be employed to the tablets. The constancy of the coating solution should be examined and explored in order to establish its shelf life.

(j) *Coating weight:* A least amount and highest amount of coating weight should be customary for the tablet. Adequate coating material has to be applied to the tablets in order to endow with a homogeneous appearance; however, it should not be huge enough to ground for fill-in of the intagliation.

(k) *Residual solvent level:* If solvents are employed for tablet coating, the residual solvent level will have to be resolute.

Exterior or appearance testing of the tablets is decisive and significant during the coating operation. Objects to be looked on for must embrace the following:

(i) Cracking or peeling of the coating.
(ii) Intagliation fill-in.
(iii) Surface roughness.
(iv) Colour uniformity.
(v) Coating competence should be determined for the coating operation. The competence will verify the quantity of coating solution overage that may be essential.

In-Process Quality Control Tests (IPQC)

In-process quality control tests are basically the regular test out that are executed during production process. They are the tests which are carried out ahead of the manufacturing process completion in order to make certain that the established product quality is ascertained before their approval for the consumption and marketing [16–18].

The purpose of in-process quality control is to monitor and if required adaptation of the manufacturing processes in order to ensure that the product conform to the specification. This includes the control of equipment and environment. In-process quality control may be carried out at usual hiatus during the process steps such as tableting or encapsulation or at the closing stages of a process step such as granulation, blending. The tests consent to the formulation scientist to categorize and pursue all alterations that transpire during applied technological procedures. It bestows the formulation scientist security that the completed products fulfil all quality prerequisites. Also, that all the products should be safe and sound for the patients [16–18].

IPQC is mainly concerned with providing of precise, explicit, and exact description of the procedures to be employed, from the reception of raw materials to the discharge of the completed dosage forms [16–18].

IPQC tests are predominantly executed within the production area.

- They should not bear any peril for the product quality.
- In-process testing enables easy recognition of problems. It sometimes makes out a substandard product batch that can be improved by re-work, but once the batch is complete, the rectification may not be possible.

- Failure to convene the in-process control specification signifies either that modus operandi was not tracked or some factor(s) were out of control.

Objectives of In-Process Quality Control

The Following Figure Depicts the Objectives of IPQC (Fig. 13.9)

- The optimization of the technological modus operandi employed in the manufacturing process.
- The monitoring, control, and effective improvement in the complete applied operations at each and every stage/step of the completed pharmaceutical products.
- The routine inspection of raw material, paraphernalia, surroundings, process, testing with special emphasis on the specification, packing, etc.
- Quality and process control.

Different IPQC tests carried out are as follows:

(a) moisture content of dried granulation
(b) granulation particle size distribution
(c) blend uniformity
(d) individual tablet/capsule weight
(e) tablet hardness
(f) tablet thickness
(g) disintegration
(h) impurity profile

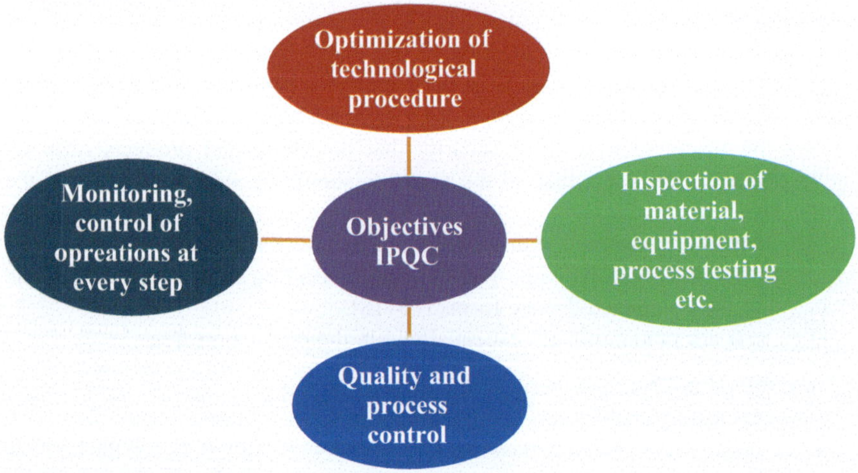

Fig. 13.9 Objectives of IPQC

The finished product tests include:

(a) appearance
(b) assay
(c) content uniformity
(d) tablet hardness
(e) tablet friability
(f) impurity profile
(g) dissolution

These vital test parameters are the index or benchmark by which the foremost processing unpredictable in solid dosage forms can be evaluated. Some processing variables are mentioned below:

(a) mixing time, blending, and granulating speed
(b) rate of solvent addition in granulators
(c) conditions like time, temperature, and airflow in dryers and coaters
(d) screen size, feed rate, and milling speed in mills
(e) machine speed and compression force in tablet presses

Note

- Process validation testing is in general done on the first three batches of product made in production size equipment.
- Revalidation testing is done only when there is a noteworthy transformation.
- A momentous change is one that will modify the in-process or final product specification created during the validation program or an alteration in formula, process, or equipment.

Change Control

Process validation of a solid dosage form must take account of a standard operating procedure (SOP) to re-examine a process whenever there are momentous and considerable alterations in the process, equipment, amenities, reactants, process materials, systems, and so on. As they have an effect on the significant quality characteristic and conditions of the solid dosage forms. Such alterations and modifications must be essayed and endorsed in accord and harmony with the extent of the change control SOP [3, 17].

The change control SOP should comprise and consist of the following fundamentals:

(a) Documentation which illustrates the procedure, review, approval, and foundation for formal revalidation studies.
(b) Detection of the change and appraisal of its likely repercussion.
(c) Requirements for monitoring alterations and testing requirements.

(d) Consideration of information and rationalization for the change.
(e) Reassess and proper approval to progress.
(f) Identification of alterations made to the physical and chemical composition of the solid dosage forms.
(g) Probable and promising regulatory action and customer notification.

Documentation

- Documentation at every phase of the process validation lifecycle is indispensable, crucial for the effectual communiqué in solid dosage form projects.
- Documentation is imperative so that acquaintance increased about a product and the process is within reach and intelligible as well as lucid to others concerned in each stage of the lifecycle.
- In totting up to being an elementary tenet of subsequent scientific method, information precision and ease of access, understanding are vital so that organizational units accountable and liable for the process can make notified, science-based verdict and assessment that will ultimately sustain the liberation of a product to commercial scale.
- The amount and category of documentation required by CGMP is maximum and supreme during process qualification, and continued process verification. Swot up throughout these stages must be conventional to CGMPs and must be endorsed and supported by the quality unit in agreement with the regulations (21 CFR 211.22 and 211.100) [6, 11–13].

Process Validation of Capsules [19]

The medication or drug mixture is contained in a hard gelatin capsule shell, a soft gelatin capsule shell, or a hard or soft shell made of any other approved material. Capsules come in a range of shapes and sizes. They typically contain a single dose of the active ingredients and are designed to be swallowed. They are basically of two types:

Hard gelatin capsules: The medication is contained in a two-part, empty hard gelatin capsule shell in this solid dosage form. The upper and little portion is referred to as the "CAP", and the remaining large portion is referred to as the "BODY". The fill volumes and eight different diameters of capsule shells (000, 00, 0, 1, 2, 3, 4) are available. The most widely used shell sizes are 0 and 2. The outer layer of hard gelatin capsules is composed of gelatin, plasticizers, and water. Modern shells can include preservatives, colours, sedatives, flavours, sweeteners, acids, enteric chemicals, and other substances.

Soft gelatin capsules: A soft gel capsule is a solid capsule with a liquid or semisolid centre, commonly known as a soft gelatin capsule (inner fill). An active substance may be present in either the outer shell or the interior filling, or both. The process for making hard gelatine capsules is the same as that for making tablets, with the exception that the granules are placed inside the capsule shell rather being compressed. The validation process is therefore the same as a result. Additional factors must be checked during the encapsulation process.

Capsule Shell Contents

Determine whether the capsule formulation is hygroscopic and whether the capsule shell is compatible with the capsule contents. Take into account the possibility that a hygroscopic formulation (API/excipients) could cause the capsule's shell to lose water, hence affecting the stability of the API.

Speed of Encapsulation

The formulation should be encapsulated at various rates in order to determine the encapsulation's operating range.

Encapsulation

Encapsulation is a crucial step in the production of capsules, much like compression is for tablet dosage forms. The materials inside must flow appropriately and be homogeneous in density.

Validation of Liquids [20, 21]

They are a category of liquid preparation where the drugs are dissolved, suspended, or distributed in an appropriate medium, and the container typically holds many doses.

Validation Includes Mainly Following Tests

- size and dispersion of particles
- particle form or morphology
- microbial count

- solvent or vehicle rheology
- solvent or vehicle pH

Monitoring Outputs

Some outputs to be monitored are as under:

- appearance
- mixing time
- pH
- viscosity
- gravity (specific gravity)
- count of microbes
- consistency of content
- testing for dissolution

The appearance of the finished product reveals signs of instability and deterioration, for instance, turbidity in an emulsion and solid particle settling in a suspension. The duration of mixing or agitation, as well as the temperature of the process, can significantly affect how something turns out. The pH of aqueous oral preparations should only be evaluated when equilibrium has been reached and at a specific temperature to minimize pH drift. It is crucial to precisely assess and validate viscosity since it affects the rate at which suspended particles settle in suspension, the coalescence of internal phase globules in emulsions, and the general look of oral solutions. A decrease in a product's specific gravity, as seen in suspensions, indicates the presence of air inside the formulation. The final product's microbial count is crucial for validation since it enables us to select the preservative for storage of the finished product. Each liquid oral solution has specific bio burden content. Drug homogeneity within the solvent system and dose uniformity are both influenced by content uniformity in multi-dose formulations.

Process Validation of Semisolid Dosages Forms [22, 23]

Creams, jellies, and pastes are examples of products that are primarily designed for external usage. Semisolids are challenging for producers to prepare because of their semisolid viscosity, which is between that of a solid and a liquid.

Critical Parameters to Be Validated

- Process temperature: Processing at the right temperature is essential for effective production. During processing, excessive heating can lead to chemical degradation, while insufficient heat can result in batch failures, and too much cooling can lead to the precipitation of solubilized components.

- Heating and cooling rates: The rate of heating and cooling, for instance, affects the consistency of ointments. If you heat too slowly, evaporative loss may cause your yields to be lower. A batch that is heated too quickly runs a higher danger of having material that has been burned when it comes into contact with the heating surface. Rapid cooling can lead to precipitation, crystallization, or increased viscosity.
- Mixing methods and speeds: It is crucial to establish the ideal mixing techniques and speeds as well as the necessary level of shear. While the mixing of a gel may require moderate shear in order to preserve certain physical qualities, such as viscosity, emulsification often requires high shear or homogenization to produce the ideal droplet size and dispersion. At every batch scale, the correct mixing speeds must be attained for each phase. The amount of shear exerted to initially scatter the polymer into the media determines how well it will hydrate. A polymer may never be fully disseminated and hydrated if the procedure only involves very low shear mixing, which could lead to an out-of-spec viscosity. Recirculation loop may be used to correct uniformity without altering mixing speed or time.
- Mixing times: To optimize mixing time, it is vital to determine the shortest amount of time needed for ingredients to dissolve and the longest amount of time before product failure (e.g. when viscosity begins to drop). The structure of polymeric gels, especially those based on acrylic acid, can be destroyed by overmixing and especially harsh shear. An emulsion can prematurely separate if it is over-mixed, which can significantly reduce viscosity.
- Flow rates: Flow rate optimization involves figuring out how much shear or throughput is needed. The addition speed may need to be slower for an oil-in-water emulsion than for a water-in-oil emulsion, for instance, thus the flow rate must be changed correspondingly. Any product that has a pump on it needs to be handled carefully. Overhearing may happen if the formulation is pumped too quickly. The formulation will spend longer time in an in-line homogenizer and be subjected to more shear if the pumping is extremely slow.
- Adding polymers and gums: When introducing polymers (Carbomers) and gums (Xanthan) directly to the mix, utmost caution must be exercised. The production of slurry of polymers or gum in a medium with little to no solubility utilizing educators like Tri-Blenders and Quadro Ytron dispersers is one method of incorporation.

Conclusion

Validation of solid dosage form should become a part of comprehensive validation programme within an industry. The multidisciplinary validation line-up must categorize the products and process characteristics that must be premeditated and integrate about specific validation tests in order to ensure that product will meet up all quality, manufacturing, and regulatory requirements.

The comprehensive validation programme should instigate with validation of the active pharmaceutical ingredients (APIs). This will serve the purpose that characteristics of the material will be standardized and consistent from batch after batch, which will provide a solid foothold upon which the dosage form will be manufactured.

Scientific in sequence acquired during the preformulation stage can form the foundation for a well premeditated and comprehensive validation programme. The parameters selected must be appropriate indicators of a controlled process. It is not adequate and satisfactory merely to work out for a test and set specifications; rather, it is enviable and advantageous to show a source and effect association between the parameter being tested and control of the quality and/or process output.

Continued attentiveness, knowledge, and understanding of validation requirements and a conscientious and meticulous application of validation principles will thus help to make certain and guarantee that pharmaceutical products will be developed and produced with the desired quality and reproducibility requisite from regulatory agencies crosswise the world.

Thus, validation is a proven assurance of the process efficiency and sturdiness and it is the full-fledged quality control tool for the pharmaceutical industries. It eliminates the chances of batch failures as the products are manufactured as per pre-optimization of each manufacturing steps. The conventional process of testing at last stage created many problems in maintaining uniformity of each batch but with the introduction of concept of validation, it has been easy to maintain the batch uniformity of the product along with imparting quality in them. This chapter summarizes the process validation stages of solids, liquids, and semisolids which are the most common pharmaceutical dosages form in use.

References

1. Aulton ME (2006) Pharmaceutics: the science of dosage form design, vol 1, 2nd edn. Churchill Livingston (Elsevier)
2. Levin M (2002) Pharmaceutical process scale-up. Marcel Dekker, Inc., New York, p 313
3. Nash RA, Wachter AH (2003) Pharmaceutical process validation, vol 129, 3rd edn. Marcel Dekker Inc, New York, pp 159–180
4. Agalloco J, Carleton FJ (2008) Validation of pharmaceutical processes, 3rd edn. Informa Healthcare USA, Inc., New York, pp 403–416
5. Lieberman HA, Lachman L, Schwartz JB (1990) Pharmaceutical dosage forms tablets, vol 3, 2nd edn. Marcel Dekker. Inc, New York, pp 117 147
6. U.S. Food and Drug Administration (1987) Guideline on general principles of process validation. U.S. FDA, Rockville, MD
7. www.wikipedia.com. Accessed 16 Aug 2020
8. Chapman KG (1991) A history of validation in the United States. Part I Pharm Tech 15(10): 82–96

9. Tetzlaff RF, Sheppard RE, LeBlanc AJ (1993) The validation story, perspectives on the systemic GMP inspection approach and validation development. Pharm Tech 17(3):100–116
10. FDA Guide on APIs, March 1998, p 48; PIC Guide, March 1999, p 32; Gold Sheet, February 1999, 6. 11
11. www.processvalidation.com. Accessed 16 Aug 2020
12. www.picscheme.org. Accessed 16 Aug 2020
13. Chaitanyakumar G, Rout RP, Ramtake S, Bhattachaiya S (2005) Process validation, The Indian pharmacist, pp 14–19
14. http://www.manufacturingtalk.com/news/nop/n op101.html. Accessed 16 Aug 2020
15. Kathiresan K, Kiran K (2005) Basics of ValidationPharmaceutical perspective, 1st edn. K.K. Publishers, Chidambaram, pp 32–46
16. http://www.pharmainfo.net/reviews/guidelinesgeneral-principles-validation-solid-liquid-and sterile-dosage-forms. Accessed 16 Aug 2020
17. http://www.pharmainfo.net/reviews/validationessential-quality-assurance-tool-pharmaindustries. Accessed 16 Aug 2020
18. http://whoguideline.blogspot.com/2008/08/imp ortance-of-master-manufacturing.html. Accessed 16 Aug 2020
19. https://www.slideshare.net/pritamkumbhar/process-validation-ofcapsules. Accessed 16 Aug 2020
20. https://www.slideshare.net/sarikakkadam/process-validation-ofliquid- orals. Accessed as 16 Aug 2020
21. International pharmacopoeia, monograph on liquid preparations for oral use, world health organization, working document QAS/07.191/rev.1
22. Idson B, Lazarus J (1991) In: Lachman L, Lieberman HA, Kanig JL (eds) Semisolids in the theory and practice of industrial pharmacy. India Varghese Publishing House, Bombay, pp 534–563
23. Nwoko P, Valentine E (2014) Semi solid dosage forms manufacturing tools, critical process parameters, strategies, optimization and validation. Sch Acad J Pharm 3(2):153–161

Part VII
Validation of Herbals

Chapter 14
Approach for Herbal Standardization

Abstract The competent scientific and ethical bodies of the individual Institutes must authorize these trials. However, when an Ayurvedic, Siddha, or Unani physician is a co-investigator in such a trial, it is essential to conduct a plant drug clinical trial. Without prior understanding of or training in any of these medical systems, any allopathic practitioner performing plant clinical trials would be acting unethically or justifiably. Therefore, it is necessary to have a representative of these systems present, and the clinical examination should be conducted collectively. In the modern era of standardizing herbal medications, pharmacognostic, chemical, biological, biopharmaceutical, and molecular methodologies to drug research and discovery are included, with biotechnology applications playing a crucial role.

Keywords Herbal drug standardization · DNA fingerprinting · Biopharmaceutical characterization · Herbal medicine · Chromatographic fingerprinting

Introduction

Folklore medicines, traditional medicines, or ethnic medicines made from plants are said to be the most ancient healthcare products ever utilized by humans worldwide. They are known to be the oldest healthcare products that have been used by mankind worldwide in the form of folklore medicines, traditional medicines, or ethnic medicines. They are the culmination of therapeutic experiences of generations of practising physicians of indigenous systems of medicine for over hundreds of years. The medicinal use of herbal medicines has seen substantial global growth during the past 10 years. According to the World Health Organization (WHO), due to its superior cultural acceptability, compatibility with the human body, and lack of negative side effects, herbal medicine is still the mainstay of primary healthcare for approximately 75–80% of the world's population, particularly in poor countries [1].

Each packet of medicine sold has the precise amount of the active component, and standardization assures that it will have the appropriate therapeutic effect [2]. Mankind has turned to nature for safer treatments as a result of the overuse of synthetic drugs that include pollutants, which has led to a higher prevalence of adverse drug reactions in more advanced cultures [2]. As a result, given the

S. S. Shukla et al., *Pharmaceutical Calibration, Validation and Qualification: A Comprehensive Approach*, https://doi.org/10.1007/978-981-19-9002-1_14

commercialization of medicinal plant-based formulations as opposed to the past, when traditional practitioners would administer the medications themselves, quality control standards for various medicinal plants used in indigenous systems of medicine are becoming more and more crucial today. There is a lot of adulteration or substitution in the commercial markets due to the various geographic places in which these plants grow as well as the problem of several vernacular names by which these plants are recognized. Therefore, each plant's reproducible standards are necessary for effective quality control [3].

A key element that can help maintain the high quality of herbal products is having sufficient control over the quality of medicinal plants. Due to natural variety, the quality of herbal starting materials generated from wild collections is becoming more unpredictable. The production of the most important medicinal plants has therefore received a great deal of attention recently. It seems to be the only way to deal with the expanding need for consistent herbal material quality in a regulated environment. As a result, the World Health Organization (WHO) has released guidelines for good agricultural and collection practices (GACP) for medicinal plants (WHO, 1991). To provide general test methods for botanical evaluation and identification of medicinal plants frequently used in conventional and home treatments, the World Health Organization (WHO) released "Guidelines for quality control procedures for medicinal plant materials" in 1992 [3].

The following technical guidelines from the World Health Organization have been released:

- The World Health Organization has provided guidelines for assessing the efficacy and safety of herbal medicines with an emphasis on contaminants and residues.
- A WHO GACP monograph is available for Artemisia Annua L.
- WHO developed recommendations for the selection of components for the quality control of herbal medicines (outline and key technical issues discussed at two WHO working group meetings in 2004 and 2005).
- WHO GMP: Supplemental guidelines have been updated for the production of herbal medicines.
- Support for building national capacity in the area of quality control for herbal medicines.
- An interregional training session on GACP and GMP for herbal medicines was funded by the WHO in China in September 2005.
- Guidelines for the Evaluation of Herbal Medicines (WHO, 1991) and Quality Control Methods for Medicinal Plant Materials (WHO, 1991).

There are guidelines for general pollution limits in addition to test methods. Standard monographs on herbs have been published by the USA, UK, ESCOP, German-E-Commission, Japanese Pharmacopoeia, and maybe the Chinese Pharmacopoeia [4].

Standardization serves diverse purposes, including:

- positive control to highlight possible loss or deterioration during manufacture
- batch to batch consistency
- confirmation of right amount of extract per dose unit

According to the American Herbal Products Association, "standardisation refers to the set of knowledge and controls necessary to manufacture material of fair consistency". To achieve this, quality assurance practices are used in manufacturing and agricultural activities to diminish the inherent variation in natural product composition [4].

For pharmaceutical purposes, the quality of medical plant material must be on par with other medicinal preparations; however, it is challenging to test for a specific chemical entity when the bioactive ingredient is unknown. Even for medicinal plant materials with active components that have been demonstrated, assay procedures are not routinely used. Similar to tea, ginseng or valerian is frequently purchased and sold based on their sensory qualities. A further issue arises from preparations involving complex heterogeneous mixtures [5].

There is a problem with standardization due to the complicated chemical makeup of medicines that are taken as entire plants, portions of plants, and plant extracts. The standardization of a drug's ostensibly active ingredients generally does not correspond to reality. In a few situations, the activity of drugs is just reliant on a single component. Most of the time, it is the result of the coordinated action of numerous active molecules along with supportive inert chemicals. It is reasonable to use the complex mixture of components given by a medicinal plant despite the inert companion components having no direct impact on the sick mechanism since they may alter the active component's absorption and excretion. Utilizing harmless plant components can also increase the stability of the active ingredient and decrease the frequency of adverse effects. There may be an additive or potentiating effect when a plant medicine has many active ingredients [5].

The Dietary Supplement Health and Education Act (DSHEA) approved by the US Congress in 1994, established the phrases "dietary supplement" and "new dietary ingredient" for herbal substances and defined the words "dietary ingredient" and "dietary supplement". A herbal ingredient may be referred to as a "new dietary ingredient" if it was not marketed in dietary supplements in the United States before October 15, 1994 [5].

The DSHEA mandates that businesses and distributors inform the FDA if they intend to market dietary supplements containing "novel" herbal ingredients. Dietary supplements containing herbal medicine are not subject to the same rigorous regulations as prescription drugs and over-the-counter medications. Under the DSHEA, dietary supplements are categorized as "foods", not "drugs", and each supplement must be labelled as such. The choice of whether or not a herbal ingredient is reasonably safe for use in dietary supplements is up to the producers and distributors [6].

Numerous international organizations and authorities, including the Department of Indian System of Medicine, the World Health Organization, the European Agency for the Evaluation of Medicinal Products and the European Scientific Cooperation in

Phytomedicine, the US Agency for Health Care Policy and Research, the European Pharmacopoeia Commission, and others, have recently started developing new mechanisms to support quality control and standardization of botanical medicine [7].

Approaches in Herbal Drug Standardization

DNA Fingerprinting

Herbal medicine requires accurate identification and quality verification of the starting material to ensure reproducible quality, which contributes to the safety and effectiveness of the treatment. Most regulatory standards and pharmacopoeias advise macroscopic and microscopic examination as well as chemical profiling of botanical materials for quality control and standardization (WHO, 1998; Indian Herbal Pharmacopoeia, 2002; British Herbal Pharmacopoeia, 1996). Finding a chemical pattern for a plant, its fractions, or extracts is known as chemical profiling. Thin layer chromatography (TLC) and high performance thin layer chromatography (HPTLC) are often used for the qualitative determination of small amounts of pollutants [8].

To achieve efficacy, the right chemotype of the plant must be selected. Even when there are several documented chemotypes of a plant species, it might be difficult to determine which chemotype is to blame for clinical consequences. Determining the identity of species that may be known by multiple binomial botanical names in various regions of the world presents another difficulty in selecting the best plant material [8–10].

A new technique that can complement or, in some circumstances, serve as an alternative is needed in light of these limitations. Molecular markers are biochemical components such as primary and secondary metabolites, as well as other macromolecules like nucleic acids. Secondary metabolites are frequently used as markers in the standardization and monitoring of the quality of herbal medicines. DNA markers are reliable for revealing polymorphisms since each species' genetic makeup is unique and unaffected by ageing, physiological conditions, or environmental factors. DNA can be extracted from fresh or dried organic plant tissue; therefore, the sample's physical form has no influence on its ability to be detected [11].

Numerous DNA-based molecular methods are used to evaluate DNA polymorphism. Hybridization, polymerase chain reaction (PCR), and sequencing are the three different sorts of procedures. Two hybridization-based strategies include restriction fragment length polymorphism (RFLP) and variable number tandem repeats. Labelled probes, including microsatellite and minisatellite probes, cDNA clones, and random genomic clones, are hybridized to filters containing DNA that has been digested by restriction enzymes. Polymorphisms are identified by the presence or lack of bands during hybridization [12].

The thermostable DNA polymerase enzyme and specific or arbitrary oligonucleotide primers are used in PCR-based markers to amplify particular DNA sequences or loci in vitro. PCR-based methods that employ random primers include random amplified polymorphic DNA (RAPD), arbitrarily primed PCR (AP-PCR), and DNA amplification fingerprinting (DAF). Amplification of genomic restriction fragments by PCR is the foundation of the amplified fragment length polymorphism (AFLP) technology. Restriction fragments are amplified using adaptor-homologous primers after being ligated with adaptors at their ends. AFLP is appropriate for DNAs of any origin or complexity and is capable of detecting hundreds of independent loci. DNA sequencing can also be used to positively identify species. It is possible to look at transversion, insertion, and deletion variations right away and get details on a particular locus [13].

Chromatographic Fingerprinting

Chromatographic fingerprinting has been employed for pharmaceutical drugs that are single chemical entities for a very long time. It has recently become one of the best ways to guarantee the quality of herbal medicines. Chromatographic fingerprinting is frequently used to identify and assess the stability of chemical compounds found in herbal medicines. Chemical and chromatographic techniques can be used to help identify a plant substance or extract. Identity tests have been carried out utilizing chromatographic methods such capillary electrophoresis, gas chromatography, thin layer chromatography, and high performance liquid chromatography (HPLC). The British Herbal Pharmacopoeia, which was published in 1996, put a lot of emphasis on using TLC profiles to characterize herbal materials and relied on them to find the active ingredients in herbal materials. Also subjected to chromatographic (HPTLC) fingerprinting were poly herbal mixtures [14].

Utilizing HPLC profiles, many ginseng varieties and sources have also been determined. In 37 samples of commercial ginseng, the authors claim to have searched for ginsenosides and malonyl ginsenosides. You must first choose the compounds you wish to quantify before you can start to build assays. If there is a known major active component, it makes the most sense to quantify it. When these chemicals are known, botanical preparations should be standardized to include the active ingredients that contribute to therapeutic efficacy [15].

A marker substance unique to the plant could be used for analytical purposes when the active ingredients are unknown; however, it should only be used for batch control. Single or more markers can be employed to confirm that the concentration and ratio of ingredients in a herbal mixture are present in reproducible amounts in raw materials, production intermediates, and final dosage forms. The information provided by many markers, often referred to as chromatographic fingerprints, helps with manufacturing control and assures batch-to-batch homogeneity [16].

Biopharmaceutical Characterization of Herbal Medicinal Products

Compared to pharmacological goods with a chemical definition, herbal medicinal products (HMPs) have less thoroughly studied biopharmaceutical quality and behaviour. In most cases, in vitro/in vivo biopharmaceutical characterization is difficult due to the complicated makeup of herbal medicine preparations, significant metabolism of components, and the accompanying analytical difficulties [17].

The active pharmaceutical ingredient (API) of HMPs is typically understood to be the entire herbal preparation, such as the extract in its entirety. Individual or groups of components have only seldom been found to be in charge of therapeutic activity [17].

There are numerous extract types that can be recognized based on pharmaceutical, analytical, pharmacological-toxicological, and clinical findings because the full herbal medicinal preparation, such as the extract, is regarded the active pharmaceutical ingredient (API). They are listed below:

- Extracts (Type A) with components (individuals or groups) wholly in charge of the well-established medicinal efficacy. It is acceptable to modify (standardize) to a specified content [18].
- Extracts (Type B1) contain chemically characterized components (single or groups) with pertinent pharmacological properties (active markers). Although there is currently no evidence that these substances are principally accountable for clinical efficacy, it is anticipated that they will add to clinical efficacy. The known efficacy, quality, and safety of an extract should be taken into consideration while characterizing it, to the extent that is practical. It is recommended to combine different lots of herbal medication preparation before extraction or to mix different lots of herbal drug preparations to standardize herbal medicine formulations. Excipients cannot be used in adjustments [18].
- Extracts (Type B2) devoid of any components that have been demonstrated to be determinants, significant for efficacy, or of pharmacological or therapeutic importance. In these circumstances, chemically defined substances (markers) with no proven therapeutic effect may be used as a control. These markers could serve as a check for good manufacturing procedures or a clue as to the drug's assay or composition [18].

This classification indicates that an extract may move from category B2 to type B1 or perhaps type A if additional information about it becomes available [18].

Relevance of the Biopharmaceutical Classification System for HMPs

HMPs might benefit from the Biopharmaceutical Classification System (BCS), which was created for chemically specified synthetic pharmacological substances [19]. In particular, a compound's solubility (in physiological pH water buffer systems) and permeability (through gastrointestinal membranes) are taken into account by the BCS.

According to the BCS, there are four categories for APIs (group I high solubility, high permeability; group II low solubility, high permeability; group III high solubility, low permeability; group IV low solubility, low permeability). High permeability is defined as more than 80% of the dose being absorbed after oral administration, whereas high solubility is defined as the highest dosage strength being soluble (> 90%) in 250 ml buffer [20]. The Note for Guidance states that the permeability of the active pharmaceutical ingredient is less significant than the solubility and rapid dissolution of the API. Scientific research suggests that permeability and solubility have an impact on an active substance's bioavailability. Pharmaceuticals, on the other hand, can only influence and regulate solubility/dissolution (e.g. by the pharmaceutical formulation) [21].

Determination of the Solubility of Extracts and Active Markers in BCS-Buffers

The greatest strength of the product's extract is tested at 37 °C to see if it dissolves in 250 ml of solvents (buffers; I-III; I: pH 1.0; II: pH 4.6; III: pH 6.8).

The leftovers are filtered and gravimetrically measured after 2 h of drying at 100 °C and 60 mins of stirring. Extracts with a solubility of >90% are regarded as very soluble since plant extracts can contain insoluble "matrix" components including tannins, proteins, and other polymeric molecules that aren't assumed to be connected to efficacy. Problematic extracts are those with a solubility of less than 90%. A difficult active marker is one whose solubility in extracts is less than 90% [22].

Based on what is known about their composition and effectiveness, plant extracts are categorized into three categories (A, B1, and B2). Herbal medicines must adhere to the Note for Guidance on Bioavailability and Bioequivalence Investigation if they contain extracts from categories A or B1 [22].

EMEA (European Agency for the Evaluation of Medicinal Products) Guidelines for Evaluation of Herbal Medicines (EMEA, 2820/00)

The term "herbal medical goods" refers to pharmaceuticals that only use herbal drugs or herbal medication preparations as its active ingredients [23].

The following evaluation procedures and standards are thought to be universal for all herbal pharmaceuticals:

- *Assay:* For products comprising herbal pharmaceuticals and/or herbal drug preparations with constituents with recognized therapeutic action, validated assays of the content of these constituents and information about the analytical technique are required (s). In the event of products containing herbal medicine(s) and/or herbal medicinal preparations, where the components responsible for the therapeutic efficacy are unknown, assays of marker chemicals, or other justified assessments are required. [23–25].
- *Impurities:* Pesticide/fumigant residues and heavy metal impurities that are found in herbal medication(s) and/or herbal drug preparations are frequently controlled during the testing of the herbal drug preparation (herbal drug), therefore testing for them in the herbal medical product is not essential. Similar to this, if the extract specification is adequately controlled, leftover solvent from the manufacture of the herbal medication preparation (such as an extract) need not be regulated in the finished herbal medicinal product. For instance, it will be necessary to keep an eye on the dose form of solvents used in tablet coating. Major pollutants brought on by the degradation of the herbal medicine preparation should be checked for in the herbal medical product (herbal drug). Products of individual degradation that may include both known and total degradation products as necessary should have acceptance limits provided [23–25].
- *Microbial limits:* It is necessary to specify the overall number of aerobic microorganisms, the total number of yeasts and moulds, and the absence of any particular harmful bacteria. These restrictions should be made in accordance with the European Pharmacopoeia. Testing needs to be supported by a case for frequency [23–25].
- *Specific tests/criteria:* On a case-by-case basis, in addition to the general tests mentioned above, herbal medical products may also be subject to the following testing. The tests should be incorporated into the batch control specification when they have an effect on the quality of the herbal medicine product. Additional tests may be needed in addition to those listed below depending on the situation or as more information becomes available [23–25].

Tablets (Coated and Uncoated) and Hard Capsules

- *Dissolution/Disintegration:* When using herbal medicines with immediate release that does not contain ingredients with known therapeutic effects, the test for in vitro active ingredient release can be skipped. For instant, release products using herbal prescription formulations that are highly soluble over the physiological pH range, disintegration tests may be sufficient. Disintegration testing is the best choice when a connection to dissolution has been demonstrated or when disintegration has been shown to be more discriminating than dissolution. Dissolution testing might not always be necessary or might be offered on a regular basis [26–29].
- Single-point readings are typically regarded as sufficient for formulations of immediate-release doses. Formulations for modified-release dosages should be tested under the proper conditions, using the appropriate sampling methods. Multiple time-point sampling is advised for extended-release dosage forms, but two-stage testing might be helpful for delayed-release dosage forms. It is crucial to take into account the populations of people or target animal species who will use the herbal medical product while developing testing and acceptability criteria (e.g. achlorhydric elderly) [26–29].
- *Hardness/Friability:* Hardness and/or friability tests as in-process controls are usually acceptable. In these situations, it is typically not necessary to provide these features in the specification. If the properties of hardness and friability have a major impact on the quality of herbal medical goods, acceptance criteria should be specified in the specification (e.g. chewable tablets) [26–29].
- *Uniformity of dosage units:* Both content and mass uniformity are covered by this expression; a pharmacopoeial protocol should be followed. If necessary, these tests can serve as in-process controls, and the specification should include the acceptance criteria [26–29].
- *Water content:* A water content test should be included when it is appropriate. The acceptance criterion may be supported by data on the effects of or water absorption on the herbal medicinal substance. In some situations, the loss on drying approach may be sufficient, but in others, a water-specific detection method (such as Karl Fischer titration) is required [26–29].
- *Microbial limits:* Microbial limit testing is regarded as both a good production practice and a means of quality control. It is not advised to test the herbal medicinal product unless all of its constituent parts have been tested before manufacture and the manufacturing process has been shown through validation studies not to represent a significant danger of microbial contamination. Consult the European Pharmacopoeia general text on the Microbiological Quality of Pharmaceutical Preparations for guidance on acceptable limits. Regular testing could be advantageous [26–29].

Chemistry-Manufacturing-Control (CMC) Considerations for Herbal Products

In contrast to normal chemically defined drugs, herbal medicines frequently had extensive human use prior to consideration in clinical trials. It is essential that the chemistry, manufacturing, and quality control of the product be the same as that of the standard formulation in order to make the most of this information in protocols for evaluating these goods [30–31]. It is not necessary to try to break down herbal medicines into their individual, known or unknown chemical components in order to evaluate herbal goods. When it comes to "analysis of the active pharmaceutical ingredient(s)" for herbal products, one or more hypothesized active ingredient(s), a chemical constituent that makes up a significant portion of the total ingredients and a chemical fingerprint of the total ingredients may be the best approaches [32–36]. The two most recent analyses serve as stand-ins for the examination of the unidentified factors that affect efficacy. The substantial issue of batch-to-batch content fluctuation in herbal products calls for the adoption of several analytical techniques to precisely measure their contents [37–39].

Conclusion

The three qualities listed below are desirable for standardization and quality assurance. Authenticity, purity, and assay are the three factors to consider. Authenticity, as the name implies, is concerned with demonstrating that the material is genuine, i.e. that it conforms to the correct identity. Gross morphology, microscopy, chemical analysis, and DNA fingerprinting are just a few of the elements that go into authentication. Purity refers to determining whether or not the plant material contains any adulterants. Chemical and biological profiling, which may examine chemical impacts and define curative values, is an assay aspect of standardization. This measure can also be used to determine the safety of a product. A pharmacological model is used to assess drug action in biological experiments. Chemoprofiling is a versatile technique that can help with standardization. In essence, fingerprinting is chemoprofiling, which is the process of creating a unique chemical pattern for a plant material, cut, fraction, or extract. Herbal medication technology is used to turn botanical materials into medicines, and it is critical to maintain uniformity and quality control while incorporating current scientific techniques and traditional expertise. The traditional methods of herbal drug standardization use botanical and organoleptic parameters of crude drugs, as well as chemoprofiling-assisted charac terization with spectroscopic techniques, but the new era of herbal drug standardization incorporates pharmacognostical, chemical, biological, biopharmaceutical, and molecular approaches.

References

1. Bhutani KK (2000) Fingerprinting of Ayurvedic drugs. Eastern Pharmacist 43:21–26
2. Blume HH, Schug BS (2000) Biopharmaceutical characterization of herbal medicinal products: are in-vivo studies necessary? Eur J Drug Metab Pharmacokinet 25:41
3. Botstein B, White RL, Skolnick M, Davis RW (1980) Construction of a genetic linkage map in man using restriction fragment length polymorphisms. Am J Hum Genet 32:314–331
4. Willoughby MJ (1996) British herbal pharmacopoeia. British Herbal Medicine Association
5. Caetano-Anolles G, Bassam BJ (1993) DNA amplification fingerprinting using arbitrary oligonucleotide primers. Appl Biochem Biotechnol 42:189–200
6. Caetano-Anolles G, Bassam BJ, Gresshof PM (1991) DNA amplification fingerprinting using very short arbitrary oligonucleotide primers. Australas Biotechnol 9:553–557
7. Chan K (2003) Some aspects of toxic contaminants in herbal medicines. Chemosphere 52: 1361–1371
8. Chau CF, Wu SH (2006) The development of regulations of Chinese herbal medicines for both medicinal and food uses. Trends Food Sci Technol 17:313–323
9. Chaudhury RR (1992) Herbal medicine for human health. World Health Organization, New Delhi, pp 51–57
10. Chauhan BL, Mitra SK, Mohan AR, Gopumadhavan S, Anturlikar SD (1994) Development of high performance thin layer chromatography fingerprint technique & bioassay to establish the self life of a proprietary herbal formulation (PHF), an Ayurvedic formulation containing multiple ingredients of herbal origin. Indian Drugs 31(7):333–338
11. ChuangW-C, Wu HK, Sheu SJ, Chiou SH, Chang HC, Chen YP (1995) A comparative study on commercial samples of *ginseng radix*. Planta Med 61:459–465
12. De Smet PAGM (2002) Drug therapy: herbal remedies. N Engl J Med 347:2046–2056
13. WHO (1991). Document No. WHO/TRM/91.4) Good agricultural and collection practices. World Health Organization, Geneva
14. WHO (1991). Document No.WHO/TRM/91.4,) Guidelines for the assessment of herbal Medicine*s*. World Health Organization, Geneva
15. Handa SS (1995) Quality control and standardization of herbal raw materials and traditional remedies. Pharmatimes 38:13–21
16. Tawi J, IDMA (eds) (1998) Indian herbal pharmacopoeia, vol I. A Joint Publication of RRL, Mumbai
17. IDMA (2002) Indian herbal pharmacopoeia. Indian Drug Manufacturers Association, Mumbai
18. Jeffrey AJ, Wilson V, Thein SL (1985) Hypervariable 'minisatellite' regions in human DNA. Nature 314:67–73
19. Joshi K, Chavan P, Warude D, Patwardhan B (2004) Molecular markers in herbal drug technology. Curr Sci 87(2):159–165
20. Joshi SP, Ranjekar PK, Gupta VS (1999) Molecular markers in plant genome analysis. Curr Sci 77:230–240
21. Kamboj VP (2000) Herbal medicine. Curr Sci 78(1):35–39
22. Kumar LS (1999) DNA markers in plant improvement: an overview. Biotechnol Adv 17:143–182
23. Lang F (2001) Biopharmaceutical characterization of herbal medicinal products. Drugs 44(4): 102–108
24. Lazarowych NJ, Pekos P (1998) Use of fingerprinting and marker compounds for identification and standardization of botanical drugs: strategies for applying pharmaceautical HPLC analysis to herbal products. Drug Inf J 32:497–512
25. Litt M, Luty JA (1989) A hypermicrosatellite revealed by in vitro amplification of a dinucleotide repeat within the cardiac muscle actin gene. Am J Hum Genet 44:397–401
26. Nakamura Y, Julier C, Wolff R, Holm TO, Connell P, Lepert M, White R (1987) Construction of a human minisatellite sequence. Nucleic Acids Res 15:2537–2547

27. Note for Guidance on specifications: test procedures and acceptance criteria for herbal drugs, herbal drug preparations and herbal medicinal products. CPMP/QWP/2820/00, EMEA/CVMP/815/00
28. WHO (1998. Document No.WHO/PHARMA/92.559/rev.1,) Quality control methods for medicinal plants materials. World Health Organization, Geneva
29. Rawls R (1996) C&EN, Washington, 23 September, 53–60
30. Savaiano J (2006) The top issues & opportunities in dietary supplements. Nutraceutical World 9(4):34–43
31. Straus SE (2002) Herbal medicines-what's in the bottle. N Engl J Med 7:1997–1998
32. Update on Eleventh ICDRA recommendations and WHO progress report, April (2006)
33. Waldesch FG, Konigswinter BS, Remagen HB (2003) Herbal medicinal products-scientific and regulatory basis for development quality assurance and marketing authorisation. Medpharm Stuttagart and CRC Press, Washington D.C., pp 37–52
34. Warude D, Chavan P, Kalpana J, Patwardhan B (2000) DNA isolation from fresh and dry plant samples with highly acidic tissue extracts. Plant Mol Biol Rep 21:1
35. Welsh J, Mcclelland M (1990) Fingerprinting genomes using PCR with arbitrary primers. Nucleic Acids Res 18:7213–7218
36. Welsh J, McClelland M (1991) Genomic fingerprints produced by PCR with consensus tRNA gene primers. Nucleic Acids Res 19:861–866
37. Willams JGK, Kubelik AR, Livak KL (1990) DNA polymorphisms amplified by arbitrary primers are useful as genetic markers. Nucleic Acids Res 18:6531–6535
38. World Health Organization (2005) Operational guidance: information needed to support clinical trials of herbal products on behalf of the special programme for research and training in tropical diseases, TDR/GEN/Guidance/05.1
39. Zabeau M (1993) European patent application., Publication No. 0534858A1

Chapter 15
Regulatory Requirements for Herbal Medicines in India

Abstract India is a significant exporter of herbs and medicinal plants. People prefer using plants and herbs as medicine since they have fewer side effects and are used in big quantities. The 1940 and 1945 D&C Acts, AYUSH, and CDSCO (Central Drugs Standard Control Organisation) all regulate the use of herbal remedies in India (amendment). Around 8000 herbal remedies have been grouped by AYUSH institutions in India. This chapter explores the regulation of herbal products in India.

Keywords AYUSH · Quality council of India · GCP · D&C Act · Clinical trials

Introduction

Medications made from herbs are used all around the world. It is a particular kind of drug that is created totally from plant materials like flowers, seeds, roots, leaves, and stems. A category of herbal medicine is botanical medicine, also referred to as phytomedicines. 80% of people currently utilize herbal medicines for some aspect of primary healthcare, according to the WHO. The broad consensus is that herbal medicines are secure and efficient. Because herbal remedies have less side effects than allopathic ones, people prefer them to allopathic ones. As referenced in the Rig-Veda and the Charaka Samhita, herbal treatments have been used in India since the time of the Vedas. According to Ayurveda, Siddha, homoeopathic medicine, and Unani, herbal medicine is practiced in India. There are numerous medicinal plants in India. Many aromatic and medicinal plants that are found in India's forests are used in the creation of medicines. There have been 8000 organized herbal therapies in India's AYUSH systems [1].

S. S. Shukla et al., *Pharmaceutical Calibration, Validation and Qualification: A Comprehensive Approach*, https://doi.org/10.1007/978-981-19-9002-1_15

Herbal Medicines

Herbal drugs/medicines are substances generated from plants that are obtained from or gathered from plants and their parts and provide therapeutic and health benefits for people. According to WHO definition, there are four categories of herbal medications:

- *Indigenous or local herbal remedies*—These herbal medicines have a lengthy history of use in a particular area, where residents have used them for therapy, composition, and dose for long periods of time.
- *Herbal medicine in systems*—Drugs in this category have been used for a longer period of time, are highly recognized in many countries, and have unique notions and theories established. They include Ayurveda, Unani, and Siddha.
- *Modified or Changed herbal medicines*—This results in modifications to herbal medicines' composition, structure, dosage form, dose, mode of administration, manufacturing processes, and medical indications. Herbal remedies must adhere to national regulatory standards in order to guarantee their efficacy and safety.
- *Imported products with an herbal drug base*—This sort of listing includes all imported herbal remedies (raw materials and finished goods). Herbal medications that are imported must be registered and sold in the country where they were made. The recipient nation's standards for safety and efficacy of regulation of herbal medicines should be addressed, and the importing nation's national authority should be supplied the medicine's safety and effectiveness data [2] (Fig. 15.1).

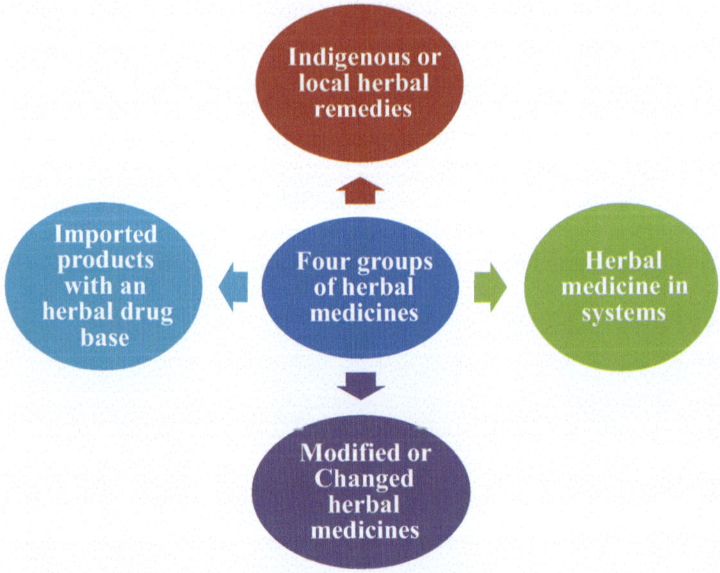

Fig. 15.1 Four groups of herbal medicines

Indian Regulations

Every known Indian health system, excluding allopathy, including ayurveda, yoga, unani, naturopathy, siddha, and homoeopathy, emphasizes the use of herbal medicines. The Central Council of Indian Medicine Act, Research Councils, the Department of AYUSH, and the D&C Act 1940 (Amendment) [2–4] all govern the use of herbal medicine in India.

The Indian Council of Medical Research, the Council of Scientific and Industrial Research, and the AYUSH department work together to create safe and efficient AYUSH products for recognized illnesses as well as to create novel pharmaceuticals in India. Herbal drugs are covered by the Drug and Cosmetics Act of 1940 and the Rules of 1945. (D & C ACT). All medications used for internal or external use, or in the diagnosis, treatment, mitigation, or cure of disease or disorder in humans or animals, and produced exclusively in accordance with the formulae described in the valid books of Siddha, the system of Ayurvedic and Unani medicine, listed in the First Schedule, are referred to as Ayurvedic, Unani, or Siddha drugs. The D&C Act covers all aspects of licencing, formulation composition, manufacture, labelling, packing, quality, and export. Regulations for Good Manufacturing Practices (GMP) are set down in Schedule "T" of the legislation for the synthesis of herbal medicines [2–4] (Table 15.1).

Ministry of AYUSH

On November 9, 2014, the Ministry of AYUSH was founded to oversee the ongoing growth and dissemination of AYUSH healthcare systems. With a focus on the advancement of education and research in Naturopathy and Yoga, Ayurvedic, Siddha, Unani, and Homeopathy (AYUSH), it was formerly known as the Department of Indian System of Medicine and Homeopathy (ISM&H), which was founded in March 1995 and renamed Department of Naturopathy and Yoga, Ayurvedic, Siddha, and Homeopathy (AYUSH) in November 2003 [5].

Objective of AYUSH

The main objective is to raise the academic standards at all Indian System of Medicine and Homoeopathy universities and colleges nationwide [5].

- To improve the state of the current research institutions and make sure that a one-time research project on illnesses that are known to exist offers an efficient treatment for these institutions.
- To design a plan for the development, regeneration, and application of medicinal plants in the aforementioned systems.

Table 15.1 Schedules for herbal products in CDSCO

Part of act/rule	Chapter/part	Nature of activity
Drugs and Cosmetics Act 1940	Chapter IV-A (Section 33-B to 33-N)	Provides guidelines for using Ayurvedic, Siddha, and Unani medicines
Drugs and Cosmetics Act 1940—schedules	First schedule	List of scheduled books
	Second schedule	Standards to be complied with by imported drugs and by drugs manufactured for sale, stocked or exhibited for sale or distributed
Drugs and Cosmetics Rules 1945	Part XVI (Rule 151–160)	Manufacture for sale of *Ayurvedic* (including *siddha*) or *Unani* drugs
	Part XVI-A (Rule 160A–160 J)	Approval of institutions for carrying out tests on ASU drugs and raw material used in their manufacture
	Part XVII (Rule 161)	Labelling, packing and limit of alcohol in ASU drugs
	Part XVII (Rule 161-B)	Shelf life and date of expiry for ASU medicines
	Part XVIII (Rule 162–167)	Government analysts and inspectors for ASU drugs
	Part XIX (Rule 168–170)	Standards of ASU drugs
Drugs and Cosmetics Rules 1945—Schedules	Schedule A	Different types of forms, particularly 24D, 24E, 25D, 25E, 26D, 26E, 26E-1, 47, 48, 49
	Schedule B-1	Fees for the test or analysis by pharmacopoeial laboratory for Indian medicine or the Govt. analyst
	Schedule E-1	List of poisonous substances under ASU systems of medicine
	Schedule FF	Standards for ophthalmic preparations
	Schedule T	Good manufacturing practices for ASU medicines
	Schedule Y	Requirements and guidelines for permission to import and/or manufacture of new drug for sale and to undertake clinical trials
	Schedule Z	Requirements and guidelines for permission to manufacture of ASU drugs for sale or for clinical trials

- To offer pharmacopoeial standards for drugs used in homoeopathy and the Indian System of Medicine [5].

By creating pharmacopoeial standards, supervising the operations of the Indian Medicine Pharmaceutical Company Limited, and keeping an eye on the Quality Council of India's pharmacopoeial laboratory, AYUSH aims to monitor the quality of medicines.

Additionally, AYUSH oversees the application of GMPs, the cluster approach to creating shared facilities, and the Drug Quality Control Scheme, which includes disclosing herbal medicine formulations, knowledge and manuscripts, documentation, and the promotion of local medical customs [5].

AYUSH Certification

There are two levels of AYUSH certification (Fig. 15.2):

- *AYUSH Standard Mark.*
  On the basis of the acquiescence to the domestic regulatory requirement.
- *AYUSH Premium Mark.*
  Based on the following options:
  - *Option A:* Based on the adherence to the GMP requirements set forth in the certification criteria and the levels of contaminants specified in the WHO guidelines [6].
  - *Option B:* Based on conformity with importing nation regulatory requirements, provided that they are stricter than option A above [6].

Drug Development Process of Herbal Medicines

The general drug development process for herbal medicines includes the following elements:

Fig. 15.2 AYUSH certification

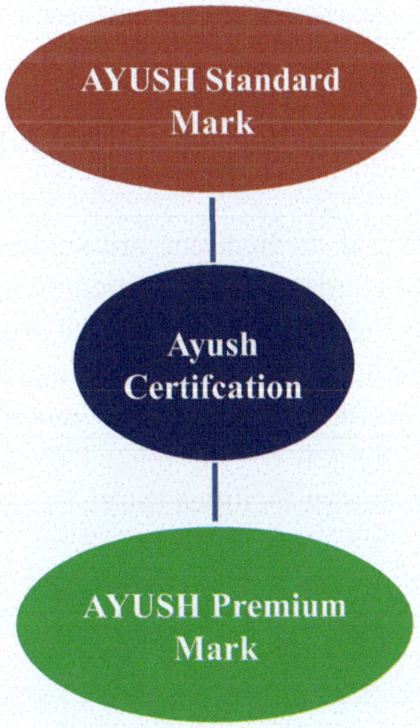

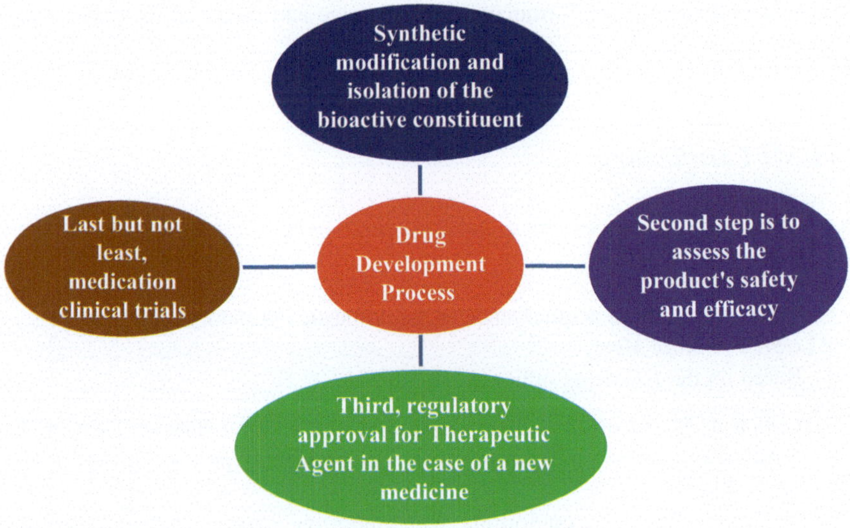

Fig. 15.3 Drug development process of herbal medicines

- The first phase is the synthetic modification and separation of the bioactive component.
- The second is the evaluation of the product's efficacy and safety.
- Third, if a new medication, regulatory clearance for the Therapeutic Agent.
- The clinical studies of medications come last but not least (Fig. 15.3).

Drug standardization is followed, among other things, by preclinical research, biological activity, and safety examinations. The standardization process for varied formulations is added where necessary [7] (Fig. 15.4).

Clinical Trials of Herbal Medicines

Herbal medications are studied and assessed on humans in accordance with the standards of Good Clinical Practice after drug testing on animals. The procedures set by the DCGI for Allopathic drugs must be followed when using herbal remedies and medicinal plants in the Allopathic System and subsequently in Allopathic hospitals. This does not apply to recommendations made by specialists in those medical systems for Ayurvedic, Unani, or Siddha drugs that can be utilized afterwards at their own facilities [8, 9].

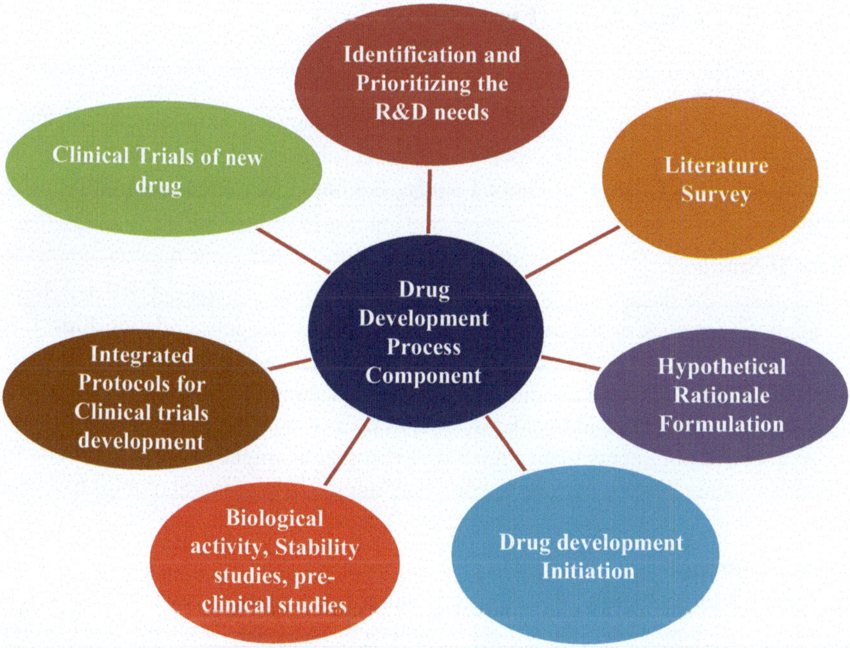

Fig. 15.4 Drug development process components

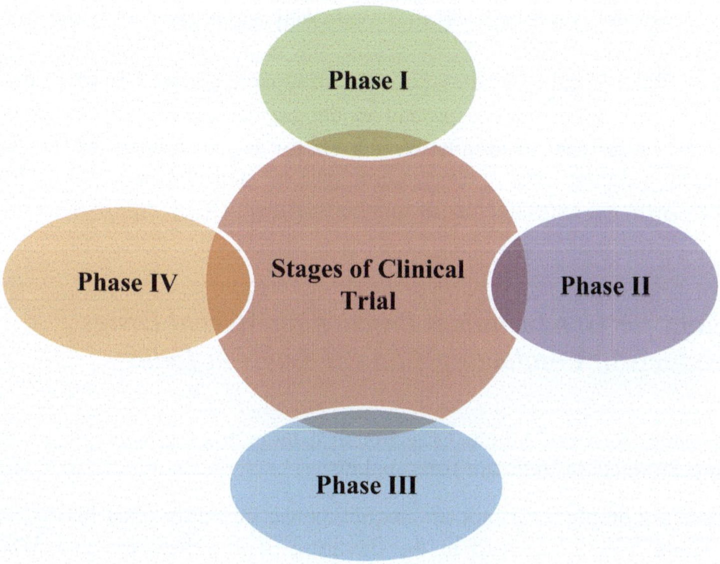

Fig. 15.5 Stages of clinical trial

Stages of Clinical Trial (Fig. 15.5)

Phase I Studies

Since they offer a reasonable level of assurance that they can be safely administered to a small number of carefully supervised clinical subjects in Phase II studies, herbal medications do not require a Phase I study. A Phase I study typically lasts many months [8, 9].

Phase II Studies

In this phase, the dosage range (100–300) for sick people is assessed. By determining ideal dose ranges or regimens and elucidating dose–response relationships, this phase aims to give a specific context for the design of major treatment trials. It is essential to test for tolerance during this stage. A comprehensive analysis of the clinical safety criteria should be the main emphasis of both the literature review and the protocol requirements. The Phase II investigation might involve hundreds of participants and takes a long time to complete, ranging from several months to a year [8, 9].

Phase III Studies

The trial, which includes extended safety and efficacy trials, is carried out after the dose-ranging Phase II data have been established. Between 1000 and 3000 people with the particular ailment are involved in clinics or hospitals throughout this phase. Patients are continually assessed to determine the efficacy of the herbal remedies and check for any negative effects. In this phase of the experiment, the drug's effectiveness and safety are confirmed. Up to 3 years may be required to finish this phase [8, 9].

In many Phase II and III studies, patients are separated into two groups and given contrasting treatments. One group receives the experimental medication, while the other receives a placebo. Additionally, these phase trials are typically conducted in the "blinded" fashion, which prevents both patients and researchers from knowing who is receiving the experimental or unique treatment [8, 9].

Guidelines for Good Clinical Practice for Herbal Drugs and Products in Conducting Clinical Trials in India

The Guidelines for Good Clinical Practice in Conducting Clinical Trials for Herbal Drugs and Products in India are listed below:

- In order to comply with industry requirements, currently used herbal remedies and plants should be listed in the literature of a well-recognized Traditional System of Medicines and processed in the same way as indicated in the literature for Good Manufacturing Practices. Phase 1 clinical trial studies are not required to be included. It is crucial to establish if the toxicity in animals has been reduced or not because the medications being examined are part of the Indian System of Medicine and are detailed in their literature. Until there is a report that suggests toxicity or when the herbal medications should be utilized, which is probably 3 months or so, toxicity testing is not necessary for the Phase II clinical research. It is crucial to do a 4- to 6-week toxicity study in the two types of animals in each of these cases [9, 10].

- Only when they have been standardized and to ensure that the indicators established for the substances being researched are consistent may clinical trials of herbal medicines be carried out. Clinical trials for plant drugs are also subject to informed consent, participants, inducements for participation, information to be delivered to the subject, withdrawal from the study, and research involving children or people with impaired autonomy. These trials must be approved by the competent scientific and ethical committees of the respective Institutes. However, it is imperative to carry out a plant drug clinical trial when an Ayurvedic, Siddha, or Unani doctor is a co-investigator in such a trial. Any allopathic practitioner conducting plant clinical trials without first having knowledge of or training in all of these medicinal systems would be acting unethically or justifying. A representative of these systems should therefore be present, and the clinical evaluation should be carried out cooperatively [9, 10].

Conclusion

Herbal remedies are employed in India's Ayurveda, Siddha, Unani, and homoeopathic medical systems. The Department of AYUSH, the ICMR, and the CSIR collaborate to create safe and effective AYUSH products for specific ailments, as well as new medications. For AYUSH medicine goods, the AYUSH department created a certification process. India creates guidelines for conducting herbal medicine clinical trials; however, the registration process is not well regulated [11, 12].

References

1. Chattopadhyay N. Maurya R. Herbal medicines [Internet]. Science Direct. https://www.sciencedirect.com/topics/agricultural-and- biological-sciences/herbal-medicines. Accessed 26 Jan 2022
2. Kumar V.(2017) Herbal medicines—overview on regulation in India and South Africa. India: WJPR. https://tinyurl.com/4bbmn5cz. Accessed 26 Jan 2022
3. Sharma S (2015) Current status of herbal products. JPBS. https://tinyurl.com/4jcsaydn. Accessed 26 Jan 2022
4. CDSCO (2020) Traditional drugs. India: CDSCO. https://cdsco.gov.in/opencms/opencms/en/Home/. Accessed 26 Jan 2022
5. Ministry of Ayush (2021). India: AYUSH. https://www.ayush.gov.in/. Accessed 26 Jan 2022

6. Voluntary certification scheme for Ayush products (2021). India: Indian Register Quality Systems (IRQS). https://www.irqs.co.in/voluntary-certification-scheme-for- ayush-products/. Accessed 26 Jan 2022
7. General guidelines for drug development of Ayurvedic formulations (2018). India: Ministry of AYUSH. https://www.ayush.gov.in/docs/guideline-drug- development.pdf. Accessed 26 Jan 2022
8. Ramadoss M S K. Koumaravelou K. (2018) Regulatory compliance of herbal medicines. IJPRS. https://pharmascope.org/ijrps/article/view/1609/2192. Accessed 26 Jan 2022
9. Good clinical practice guidelines. India: Department of AYUSH; 2013. https://main.ayush.gov.in/sites/default/files/5110899178- Final%20Book%2028-03-13_0.pdf. Accessed 26 Jan 2022
10. Srivastava A. Jain R. Agrawal R.K. Ahirwar D (2008). Clinical trials of herbal products in India. India: RJPT. https://tinyurl.com/6rp7m2ca. Accessed 26 Jan 2022
11. Chegu S, Nagabhushanam MV (2021) A comprehensive study on regulation of herbal drugs in India, US and European Union. Int J Drug Reg Affairs 9(1):78–76. http://ijdra.com/index.php/journal/article/view/458. Accessed 26 Jan 2022
12. Budhwar V, Yadav S, Choudhary M, Nitesh (2014) A comprehension study on regulation of herbal drugs in USA, European Union and India. Int J Drug Reg Affairs 5(4):8–17. http://ijdra.com/index.php/journal/article/view/205. Accessed 26 Jan 2022

Part VIII
Expert Opinion

Chapter 16
Expert Opinion on the Basis of Table of Contents

Calibration is a critical component of any organization's performance and growth in the pharmaceutical sector and laboratories. Instrument, equipment, process, and related services calibration should be pre-planned and scheduled according to a defined protocol. In the calibration process, accuracy and dependability are critical. For a calibration laboratory, there are a few basic requirements that must be met. Calibration is always done in a controlled environment in the I/II/III party laboratory. (Only to the extent necessary, the environment must be managed.) It should be carried out at an accredited calibration laboratory, as well as a manufacturer's or supplier's laboratory, to ensure accurate traceability. Inspection, measuring, and testing of equipment are all common measurement characteristics (IM&TE). To calibrate any equipment, a known amount of the variable to be measured must be generated and applied to the unit under test. The four essential factors that must be calibrated in any equipment are temperature, humidity, pressure, and mass. The frequency of calibration, which includes recalibration, is also a requirement. As a result, calibration is critical wherever quantitative measurements are required, as it ensures accuracy.

Uncertainty analysis is the process of finding and qualifying errors. Two essential aspects that contribute to certainty estimation are accuracy and precision. To avoid concerns of uncertainty, the idea of uncertainty must be used in the validation of analytical methods. However, there isn't a well-established process for estimating calibration uncertainty. In the future, investigating uncertainty to achieve precision and reliability will be vital and crucial aspects in any process.

All measurement tools are inaccurate, and calibration determines how inaccurate they are. Calibration has become a need in everyday life for achieving and obtaining the highest performance in product manufacturing, production, and output. A calibration curve is a mathematical function relationship between observed responses and known values of reference standards. Equipment should be calibrated before use, after use, and whenever necessary; however, if calibration is not required, some equipment may be used without calibration. In its most basic form, calibration is the process of determining an instrument's accuracy. Special attention to the design of

S. S. Shukla et al., *Pharmaceutical Calibration, Validation and Qualification: A Comprehensive Approach*, https://doi.org/10.1007/978-981-19-9002-1_16

the equipment/instrument to be calibrated, checking the tolerance, maintaining the accuracy ratio, and adhering to standards are the primary requirements for the calibration process. So that difficulties do not arise during regulatory inspections, an authorized lab AS 17025 or ISO 17025 can give the proper calibration certifications for each piece of equipment. Failure to produce valid paperwork is a violation of FDA laws and pharmaceutical cGMP, and failure to do so can result in penalties, fines, and even product recalls. The calibration procedure varies depending on whether the instrument is linear, nonlinear, or discrete. Uncertainty must also be measured, which comes from three basic sources. This will give the equipment a fresher look while also extending its life and improving its performance.

Instruments, equipment, and related services are used by pharmaceutical manufacturing enterprises to manufacture high-quality products with the greatest efficiency. All plant instruments and measurement devices are identified and categorized as "critical" or "non-critical". Measurement equipment, sample detection, operation record, certificate life cycle management, multichannel customer notification, and intelligent early warning are all areas where the calibration management information system excels. It ensures a higher level of refined management, standardization of corporate processes, increased work productivity, ongoing quality management, and better customer service. In the future, software calibration will be required in every business to preserve procedure and obtain the best potential results.

Qualification is a word that refers to the process of transferring equipment from a vendor to an owner's site, starting with planning, executing, and obtaining the findings of an inquiry or analysis in order to check the equipment's capacity to work. The equipment must meet the acceptance requirements provided in the supplier's design qualification specification and guidelines. Qualification shall be completed on regular intervals and is relevant for the equipment, instruments, facility, and area before its use. There are several stages to the certification process, including the specification phase, which includes user requirement specifications (URS), factory acceptance tests (FAT), site acceptance tests (SAT), and design qualification. Installation qualification, operational qualification, and performance qualification are all part of the second phase verification. All new/existing equipment, facility, system, and instrument shall be qualified by the user with the assistance of the manufacturer/supplier of the equipment, instrument, system, and facility, as well as a qualified qualification team and engineering person. Validation, qualification, and calibration are all interrelated phrases that refer to a group of operations rather than a single event. If a facility's development, operation, and maintenance are to continue to meet all regulatory standards, these three must be done on a regular basis. Qualification is essential since it contributes to the system's quality and performance.

In the pharmaceutical business and laboratories, qualification is a key aspect in the selection, procurement, installation, operation, and performance of any equipment, from the manufacturer to the vendor. It is the complete process of obtaining items from manufacturers or distributors, inspecting and testing them, and finally identifying them as qualified products. Nature (who, what, (potentially where), why, and how is the action being objectively proven) and time (when can the activity begin and when must the activity conclude) are critical for qualifying execution and implementation, as covered in this chapter. The regulatory agencies FDA, GMP, and ICH will emphasize the mandatory use of techniques for certification in the future.

Verification, in simple terms, is the process of confirming the accuracy of a hypothesis or fact. The act of reviewing, inspecting, or testing a product, service, or system in order to establish and record that it fulfils regulatory or technical requirements is known as verification and validation in engineering and quality management systems. Static verification is focused with analysing the static system representation in order to find issues, whereas software testing is concerned with exercising and observing product behaviour as an indicator of dynamic verification. Planning, execution, and reporting are the three processes in the verification process. Testing, static verification, inspection/review/walkthrough, and formal proof are examples of verification methods. Every firm should employ verification as a crucial procedure to assure system dependability and reproducibility. Validation and verification work in tandem and are inextricably linked. The quality of software can be improved by employing the V&V process to verify and validate it on a regular basis. Verification is critical to a software system's operation, accuracy, and output outcomes. Defect testing and statistical testing are two types of verification testing that are commonly used to estimate dependability. Verification and validation should provide assurance that the programme is fit for its intended use. Defect testing and debugging are two different things. The purpose of verification and validation is to determine whether or not a programme has any faults. Debugging is the process of detecting and correcting mistakes. The "Clean room" technique, which was developed with the idea of avoiding flaws rather than removing them, is the most recent contribution to software verification. More of these emerging technologies may help save time and money by improving efficiency and reproducibility, as well as providing long-term features for software verification.

The continuous process improvement in software testing not only ensures higher product quality but also optimizes business processes. Software quality is the degree to which the correct software produced. Quality software is reasonably bug or defects free, delivered on time and within budget, meets requirements, expectations, and maintainable. Software life cycle testing essentially means that testing occurs parallelly with the development cycle and is a continuous process. It is important to start the software testing process early in the application life cycle, and it should be integrated into application development itself. Quality control and quality assurance play a key role in improving the quality of software.

Validation is an important step in establishing and maintaining the quality of the final product. If each step of the production process is examined, we can ensure that the end product is of the highest quality. Validation is the skill of generating and practising the processes as well as the documentation that have been designed. In the disciplines of medicine research, manufacturing, and finished product specification, the term "validation" is most typically employed. Consistency and reliability of a proven process to offer a quality product are crucial for an industry.

Process validation assures product quality, uniformity, and reproducibility, as well as the safety of pharmaceutical products, as mandated by regulatory bodies all over the world. To ensure that the product meets all quality, manufacturing, and regulatory criteria, the multidisciplinary validation team must first identify, evaluate, and incorporate the crucial needed validation key parameters. The most significant and well-known cGMP parameter is process validation. The process validation is meant to help producers understand the needs of their quality management system (QMS) for process validation, and it may be applied to any manufacturing process.

Cleaning validation, in conclusion, is the process of obtaining and documenting adequate evidence to demonstrate the success of a cleaning method. Cleaning is closely related to the pharmaceutical product's safety and purity; thus, it becomes the most critical and primary task. As a result, the regulatory requirement necessitates the implementation of an effective cleaning programme. This chapter goes through everything you need to know about cleaning validation, including residue selection, validation acceptance criteria, different levels of cleaning, cleaning technique, sample procedure, product grouping and equipment characterization, and cleaning chemical selection.

Validation of solid dosage form should become a part of comprehensive validation programme within an industry. The multidisciplinary validation line-up must categorize the products and process characteristics that must be premeditated and integrate about specific validation tests in order to ensure that product will meet up all quality, manufacturing, and regulatory requirements.

The comprehensive validation programme should instigate with validation of the active pharmaceutical ingredients (APIs). This will serve the purpose that characteristics of the material will be standardized and consistent from batch after batch, which will provide a solid foothold upon which the dosage form will be manufactured.

Scientific in sequence acquired during the preformulation stage can form the foundation for a well-premeditated and comprehensive validation programme. The parameters selected must be appropriate indicators of a controlled process. It is not adequate and satisfactory merely to work out for a test and set specifications; rather, it is enviable and advantageous to show a source and effect association between the parameter being tested and control of the quality and/or process output.

Continued attentiveness, knowledge, and understanding of validation requirements and a conscientious and meticulous application of validation principles will thus help to make certain and guarantee that pharmaceutical products will be developed and produced with the desired quality and reproducibility requisite from regulatory agencies crosswise the world.

Thus, validation is a proven assurance of the process efficiency and sturdiness, and it is the full-fledged quality control tool for the pharmaceutical industries. It eliminates the chances of batch failures as the products are manufactured as per pre-optimization of each manufacturing steps. The conventional process of testing at last stage created many problems in maintaining uniformity of each batch but with the introduction of concept of validation, it has been easy to maintain the batch uniformity of the product along with imparting quality in them.

The three qualities listed desirable for standardization and quality assurance are authenticity, purity, and assay the three factors to consider. Authenticity, as the name implies, is concerned with demonstrating that the material is genuine, i.e. that it conforms to the correct identity. Gross morphology, microscopy, chemical analysis, and DNA fingerprinting are just a few of the elements that go into authentication.

Purity refers to determining whether or not the plant material contains any adulterants. Chemical and biological profiling, which may examine chemical impacts and define curative values, is an assay aspect of standardization. This measure can also be used to determine the safety of a product.

A pharmacological model is used to assess drug action in biological experiments. Chemoprofiling is a versatile technique that can help with standardization. In essence, fingerprinting is chemoprofiling, which is the process of creating a unique chemical pattern for a plant material, cut, fraction, or extract.

Herbal medication technology is used to turn botanical materials into medicines, and it is critical to maintain uniformity and quality control while incorporating current scientific techniques and traditional expertise. The traditional methods of herbal drug standardization use botanical and organoleptic parameters of crude drugs, as well as chemoprofiling-assisted characterization with spectroscopic techniques, but the new era of herbal drug standardization incorporates pharmacognostical, chemical, biological, biopharmaceutical, and molecular approaches.

Herbal remedies are employed in India's Ayurveda, Siddha, Unani, and Homoeopathic medical systems. The Department of AYUSH, the ICMR, and the CSIR collaborate to create safe and effective AYUSH products for specific ailments, as well as new medications. For AYUSH medicine goods, the AYUSH department created a certification process. India creates guidelines for conducting herbal medicine clinical trials; however, the registration process is not well regulated.